HEAT TRANSFER IN AEROSPACE APPLICATIONS

HEAT TRANSFER IN AEROSPACE APPLICATIONS

BENGT SUNDÉN

JUAN FU

Amsterdam • Boston • Heidelberg • London
New York • Oxford • Paris • San Diego
San Francisco • Singapore • Sydney • Tokyo
Academic Press is an imprint of Elsevier

Academic Press is an imprint of Elsevier
125 London Wall, London EC2Y 5AS, United Kingdom
525 B Street, Suite 1800, San Diego, CA 92101-4495, United States
50 Hampshire Street, 5th Floor, Cambridge, MA 02139, United States
The Boulevard, Langford Lane, Kidlington, Oxford OX5 1GB, United Kingdom

Notices
Knowledge and best practice in this field are constantly changing. As new research and
experience broaden our understanding, changes in research methods, professional
practices, or medical treatment may become necessary.

Practitioners and researchers must always rely on their own experience and knowledge in
evaluating and using any information, methods, compounds, or experiments described
herein. In using such information or methods they should be mindful of their own safety
and the safety of others, including parties for whom they have a professional
responsibility.

To the fullest extent of the law, neither the Publisher nor the authors, contributors, or
editors, assume any liability for any injury and/or damage to persons or property as a
matter of products liability, negligence or otherwise, or from any use or operation of any
methods, products, instructions, or ideas contained in the material herein.

Library of Congress Cataloging-in-Publication Data
A catalog record for this book is available from the Library of Congress

British Library Cataloguing in Publication Data
A catalogue record for this book is available from the British Library

ISBN: 978-0-12-809760-1

For information on all Academic Press publications
visit our website at https://www.elsevier.com/

Working together
to grow libraries in
developing countries

www.elsevier.com • www.bookaid.org

Publisher: Joe Hayton
Acquisition Editor: Carrie Bolger
Editorial Project Manager: Carrie Bolger
Production Project Manager: Mohana Natarajan
Designer: Mark Rogers

Typeset by TNQ Books and Journals

CONTENTS

PREFACE

The requirements of thermal management in aerospace applications are continuously growing, whereas the allotments on weight and volume remain constant or shrink. To meet the high heat flux removal requirements, compact, high-performance, and lightweight heat transfer equipment are needed. Heat exchangers based on microchannels are very suitable, as they offer opportunities for high heat flux removal because of their good thermal performance and extremely compact size. However, aerospace challenges include reduced gravity or microgravity, low or no atmospheric pressure, extreme temperatures, aerodynamic heating, dynamic vibration, shock loads, and extended duration of operations. Also alternative power sources are needed for aerospace vehicles, e.g., fuel cells. As hydrogen is the common fuel, efforts have been spent on its production, transportation, storage, system design, and safe and effective handling. Heat transfer issues are also demanding challenges for aerospace propulsion. It is important to protect the propulsion surfaces from the hostile thermal environment. One way to achieve this is to develop materials capable of withstanding the hostile environment and offering an adiabatic surface that will not melt or lose its structural integrity. Another approach is to immediately cool the exposed surfaces. Heat transfer issues in hypersonic flights include very high aerodynamic loads, laminar–turbulent transition, shock/shock and shock/boundary layer interactions, film cooling and skin friction reduction, advanced composite materials, combined thermal/structural analysis, real-gas effects, and wall catalysis, as well as thermal management of the integrated engine–airframe environment. Heat pipes are potential candidates for passive cooling of structures exposed to very high heat flux levels. It is obvious that heat transfer engineering and thermal sciences are important for the design and development in aerospace applications.

The idea to write this book was created after the senior author presented a short lecture series on aerospace heat transfer issues at the National University of Defense Technology, Changsha, Hunan, China in 2013. During the preparation stages, it was found that the textbook treated the topics of heat transfer in aerospace applications sufficiently well.

Some coworkers were quite helpful in the writing of this book and preparation of figures, as well as searching published literature. They are Dr. Zan Wu, Dr. Chenglong Wang, Dr. Luan Huibao, and Dr. Daniel Eriksson. Ms. Carrie Bolger and Ms. Mohana Natarajan at Elsevier were also quite helpful in bringing this book to completion.

Lund and Beijing, July 2016
Bengt Sundén and Juan Fu

NOMENCLATURE

A	Area (m^2)
A	Reaction activation energy
a	Velocity of sound (m/s)
Bo	Bond number
b	Thickness (m)
C	Specific heat matrix
C_f	Convective flux (kg/s)
c_p	Specific heat constant pressure [J/(kg K)]
c_v	Specific heat constant volume [J/(kg K)]
c	Specific heat [J/(kg K)]
c_D	Drag coefficient
D	Diameter (m)
D_f	Diffusive flux (kg/s)
d	Distance (m)
d_{ij}	Deviatoric stress tensor (N/m^2)
$E, \dot{E}$	Energy rate (W)
E	Young's modulus (Pa)
e	Internal energy (J/kg)
e_{ij}	Rate of strain tensor (1/s)
F	Fill level
$F_i, \overrightarrow{F}$	Body force (N/m^3)
F_{12}	View factor
f_s	Accommodation coefficient
f	Velocity function
Gr	Grashof number
G_s	Solar radiation (W/m^2)
g	Gravity acceleration (m/s^2)
H	Enthalpy [J/(kg K)]
h	Heat transfer coefficient [W/(m^2 K)]
h	Enthalpy (J/kg)
K	Permeability
Ka	Kapitza number
K_c	Conductivity matrix
Kn	Knudsen number
k	Thermal conductivity [W/(m K)]
k	Boltzmann constant
L	Length (m)
L_H	Latent heat (J/kg)
l	Characteristic length (m)
Ma	Mach number, U/a
$\dot{m}$	Phase change mass flow rate (kg/s)

N	Mesh number
Nu	Nusselt number
n	Integer
n	Number of molecules per unit volume
n	Normal vector
Pr	Prandtl number
p, P	Pressure (Pa)
p	Width (m)
$Q, \dot{Q}$	Heat transfer rate (W)
Q	Heat flow vector
Q_L	Latent heat (J/kg)
q	Heat flux (W/m^2)
R	Gas constant [J/(kg K)]
Ra	Rayleigh number
Re	Reynolds number
r	Recovery factor
r	Radial coordinate (m)
r	Adjustable coefficient
r_1, r_2	Principal radii of curvature (m)
S	Molecular velocity ratio
S_h	Source term energy equation (W/m^3)
St	Stanton number
s	Pitch (m)
T	Temperature (°C, K)
T_{aw}	Adiabatic wall temperature (°C, K)
T^*	Reference temperature (°C, K)
t	Temperature (°C)
t_B, t_w, t_T	Thicknesses (m)
U	Velocity (m/s)
u	Velocity (m/s)
V	Average or absolute velocity (m/s)
V_g	Vapor volume (m^3)
V_t	Tank volume (m^3)
$\overrightarrow{V}$	Velocity vector (m/s)
$\overline{v}$	Molecular mean velocity (m/s)
u, v, w	Local velocity (m/s)
v_1, v_2, v_3	Molecular velocity component (m/s)
$W, \dot{W}$	Work rate (W)
We	Weber number
X	Thickness (m)
x, y, z	Coordinates (m)

Greek

α	Heat transfer coefficient (W/(m^{2}K)]
α	Thermal diffusivity (m^2/s)
α	Thermal expansion coefficient (1/K)
α	Volume fraction
α_s	Absorptance
α_s	Accommodation coefficient
β	Inclination angle
β	Thermal expansion coefficient
Δ	Rate of expansion (1/s)
Δp	Pressure drop (Pa)
δ	Boundary layer thickness (m)
δ	Depth (m)
δ_{ij}	Kronecker's delta
ε	Emissivity
ε	Porosity
ϕ	Arbitrary variable
ζ	Dimensionless parameter
γ	c_p/c_v
κ	Curvature (1/m)
η	Dimensionless coordinate
θ	Dimensionless temperature
θ	Polar angle, angle
θ	Temperature difference
ϑ	Dimensionless temperature
λ	Mean free path (m)
λ	Dimensionless parameter
μ	Dynamic viscosity [kg/(m s)]
μ	Poisson's ratio
ν	Kinematic viscosity (m^2/s)
ν	Dimensionless parameter
ρ	Density (kg/m^3)
σ	Surface tension (N/m)
σ	Stefan–Boltzmann constant [W/(m^{2}K^4)]
σ_s	Accommodation coefficient
τ	Time (s)
ψ	Stream function (m^2/s)
ψ	Angle
Ω	Dimensionless parameter
ω	Length

Indices

a	Adiabatic
aw	Adiabatic wall
b	Bulk
c	Continuum
eff	Effective
fm	Free molecule
l	Liquid
lv	Liquid−vapor
tj	Temperature jump
v	Vapor
vol	Volume
w	Wall
∞	Free stream

Abbreviations

ACC	Advanced carbon-carbon
BEM	Boundary element method
CAD	Computer-aided design
CFD	Computational fluid dynamics
CMC	Ceramic matrix composites
CMG	Compression mass gauge
CVFEM	Control volume finite element method
CO_2	Carbon dioxide
DNS	Direct numerical simulation
DSMC	Direct simulation Monte Carlo
ECS	Environmental control system
FEM	Finite element method
FVM	Finite volume method
H_2	Hydrogen
H_2O_2	Hydrogen peroxide
Ir	Iridium
LH_2	Liquid hydrogen
LOX	Liquid oxygen
MMC	Metallic matrix composites
μHEX	Micro heat exchanger
$NaBH_4$	Sodium borohydride
NB	Neighbor
O_2	Oxygen
PCHE	Printed circuit heat exchanger
PFHE	Plate-fin heat exchanger

PISO	Pressure implicit splitting operators
QUICK	Quadratic upstream interpolation for convective kinetics
RANS	Reynolds-averaged Navier—Stokes
RSM	Reynolds stress method
SiC	Silicon carbide
SIMPLE	Semi-implicit method for pressure-linked equations
SIMPLEX	SIMPLE extended
SIMPLER	SIMPLE revised
SiO	Silicon oxide
SST	Shear stress transport
TDMA	Tridiagonal matrix algorithm
TPS	Thermal protection system
VOF	Volume of fluid
Zr	Zirconium

CHAPTER 1

Introduction

1.1 HEAT TRANSFER IN GENERAL

Heat is a form of energy that is always transferred from the hot part to the cold part in a substance or from a body at a high temperature to another body at a lower temperature. The bodies do not need to be in contact but a difference in temperature must exist.

In some cases the amount of heat transferred can be determined simply by applying basic relations or the laws of thermodynamics and fluid mechanics. In other cases in which the mechanisms of heat transport are not completely known, methods of analogy or empirical methods based on experiments are applied.

Heat can be transferred by three different means, namely, heat conduction, convection, and thermal radiation, as illustrated in Fig. 1.1. Many textbooks are available on general heat transfer, see, e.g., Refs. [1–3].

Heat conduction is a process in which energy transfer from a high temperature region to a low-temperature region is governed by the molecular motion, as in solid bodies and fluids (gases and liquids) at rest, and by the movement of electrons, as for metals.

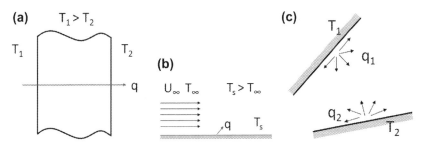

Figure 1.1 Heat transfer by (a) heat conduction, (b) convection, and (c) thermal radiation.

Heat Transfer in Aerospace Applications
ISBN 978-0-12-809760-1
http://dx.doi.org/10.1016/B978-0-12-809760-1.00001-6

For heat conduction across the wall in Fig. 1.1 the heat flux is calculated by

$$q = k(T_1 - T_2)/b \qquad (1.1)$$

$$Q = q \cdot A \qquad (1.2)$$

where q is the heat flux in W/m^2; k, the thermal conductivity of the wall material in W/mK; b, the wall thickness; T_1 and T_2, the temperatures on the wall surfaces; and Q, the total amount of heat transferred in W.

When a fluid is flowing along an exterior surface or inside a duct and if the temperatures of the fluid and the solid surface are different, the amount of heat being exchanged is affected by the macroscopic fluid motion. This type of heat transfer is called convection. Depending on how the macroscopic fluid movement is created, forced convection or free (natural) convection prevails. In some cases, both forced and free convection occur simultaneously. The process is then called mixed convection or combined forced and free convection. For this heat transfer mode a heat transfer coefficient is introduced according to

$$q = \alpha(T_S - T_\infty) = h(T_S - T_\infty) \qquad (1.3)$$

where α and h is the heat transfer coefficient in W/m^2K and T_S and T_∞ are the temperatures of the surface and fluid, respectively. The heat transfer coefficient is, in general, a function of flow velocity, fluid type, temperature, and geometry.

Heat transfer by radiation does not require any medium to propagate. The heat transfer between two surfaces by radiation is in fact maximum when no media is present between the surfaces. Radiation may occur between surfaces, as well as between a surface and a participating medium, such as gas. Heat exchange by radiation is governed by electromagnetic waves according to Maxwell's theory or in the form of discrete photons according to Planck's hypothesis. For the solid body radiative heat exchange illustrated in Fig. 1.1, the heat flow rate is given by

$$Q_{net} = A_1 F_{12} \varepsilon_{eff} \left(T_1^4 - T_2^4\right) \qquad (1.4)$$

where F_{12} is the view factor; ε_{eff}, the effective emissivity; and A_1, the area of surface 1.

If a medium participating in radiation is present between the surfaces, the calculation of the radiative heat exchange becomes more complicated [4,5].

In many branches of engineering and technology, it is of great interest to be able to calculate temperature distributions and heat fluxes. In order to design, size, and rate heat exchangers, e.g., condensers, evaporators, and radiators, analysis of heat transfer is needed. Huge applications of this type of equipment appear frequently in heat and power generation, process industries, etc. Design and sizing of air-conditioning equipment, electronics cooling, and thermal insulation of buildings require knowledge in heat transfer. For vehicles, many heat transfer problems are present.

To enable stress and strain analysis in equipment exposed to high temperature and/or gradients, analysis of the temperature field and heat loads is needed. In manufacturing, production, and treatment of materials, heat transfer is also important. Cooling of electronics and other equipment carrying electric currents is an important application area of heat transfer. Also in combustion devices, heat transfer is of significance because of thermal radiation and convection. Processing and treatment of food require analysis of heat and mass transfer.

1.2 SPECIFICS FOR AEROSPACE HEAT TRANSFER

Thermal management requirements for aerospace applications grow continuously, whereas the allotments on weight and volume remain constant or shrink. To meet the high heat flux removal requirements, compact, high-performance, and lightweight heat transfer equipment are needed. Heat transfer systems in military aircraft are increasingly using fuel as a heat sink. Heat transfer loops involving several fuel-to-liquid heat exchangers are used to cool electronics, engine oil, hydraulic oil, and elements of the thermal management system. Heat exchangers based on microchannels are very suitable, as they offer opportunities for high heat flux removal because of their good thermal performance and extremely compact size.

Aerospace challenges include reduced gravity, low or no atmospheric pressure, extreme temperatures, aerodynamic heating, dynamic vibration, shock loads, and extended duration operations. Also alternative power sources are needed for aerospace vehicles. One of the possible alternatives is fuel cells. Providing reliability, compactness, and high-energy power sources for aerospace applications is important, and fuel cells might be a good candidate. As hydrogen is a common fuel, a lot of effort has been spent on its production, transportation, storage, system design, and safe and effective handling.

Heat transfer issues are also demanding challenges for aerospace propulsion. In general, the chemical energy in the fuel is transformed into useful work of propulsive thrust at maximum effectiveness. The propulsion system must then operate at a very high temperature and pressure. It is important to protect the propulsion surfaces from the hostile thermal environment. One way to achieve this is to develop materials capable of withstanding the hostile environment and offer an adiabatic surface that will not melt or lose its structural integrity. Another approach is to immediately cool the exposed surfaces.

For subsonic and supersonic flights, the turbine engine is the backbone. The hot sections include the combustor, the turbine, the exhaust valves, and some other components. The turbine is deemed to be the most demanding, and heat transfer challenges are associated with proper treatment of unsteadiness, turbulence modeling, film cooling, complex internal flow passages with various surface modifications, rotational effects, new materials, and thermal stress analysis. Film cooling is important for the turbine, combustor, and nozzles. It has been the primary method to protect the hot sections of the turbine engine from the hostile thermal environment. However, traditionally film cooling design has been based on empirical correlations obtained from specific experiments concerning both the shape of the holes and their positions. The turbine airfoil has a complex internal serpentine flow passage with ribs or other protrusions on the walls; impinging jets, pin fins, and film cooling are also present. The passages are generally short and entrance effects might be significant. The passages are also subject to rotation and accordingly, buoyancy forces and Coriolis forces occur.

Space propulsion is by rocket, and many space missions rely on liquid hydrogen and liquid oxygen propellants because of their high specific impulse. However, there are storage problems on and near the launchpad and in space. Space storage and transfer of cryogenic propellants near the launchpad and on the space station pose unusual heat transfer and material problems. Space missions require large propellant storage depots, and for some missions, propellant production on the planets or their moons must be given consideration. For subcritical propellant storage on the Earth, one usually knows where the liquid is, but in space or bases with microgravity the propellant orientation is a serious concern. In addition, heat leakage from support structures and fittings and insulation and degradation caused by impalement from space debris represent challenges to long-term storage of cryogens in space. In microgravity, local nonuniform sources of heat usually disturb the average void/liquid fraction, which, along with the energy of the container surface, determines how the liquid will be distributed.

Flight at hypersonic speeds presents important heat transfer challenges throughout the propulsion system and the airframe because of the very high aerodynamic heat loads encountered. Local stagnation areas can experience very high heat fluxes exceeding $50 \, kW/cm^2$. The list of heat transfer issues at hypersonic flights includes very high aerodynamic loads, laminar—turbulent transition, shock/shock and shock/boundary layer interactions, film cooling and skin friction reduction, advanced composite materials, combined thermal/structural analysis, real gas effects, and wall catalysis, as well as thermal management of the integrated engine—airframe environment. Heat pipes are potential candidates for passive cooling of the structures exposed to very high heat fluxes.

1.2.1 Thermal Management

There are several options for handling the severe thermal environments encountered during supersonic and hypersonic flights. Passive and semipassive and actively cooled methods can be utilized. The differences between rockets and air breathers can have a significant impact on the thermal protection system (TPS) and the hot structures. The passive and semipassive thermal management methods may include a phase change, whereas the actively cooled structures may include a pumped coolant.

Fig. 1.2 illustrates an insulated structure applicable for moderate heat fluxes over relatively short periods. The surface is heated and thermal radiation is the mechanism to remove the heat. The insulation should minimize the heat reaching the structure so that it can remain cool. A small amount of heat is conducted through the insulation to the structure.

In Fig. 1.3 a heat sink structure is shown. It can be used for moderate heat fluxes and under transient conditions. As the surface is being heated, part of the heat is radiated away and the rest is absorbed by the structure. However, for long exposition times, there is a risk of overheating of the structure.

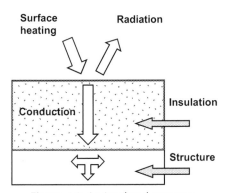

Figure 1.2 An insulated structure.

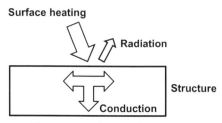

Figure 1.3 A heat sink structure.

Hot structures are utilized for moderate heat fluxes and for long periods. The hot structure is exposed to high heat loads for long periods, and the structure is allowed to reach a steady-state condition. The heat transfer mechanisms are radiation and conduction. An illustration is provided in Fig. 1.4.

If a high heat flux prevails over a long period, a semipassive cooling technique may be used. A heat pipe might be used for high heat fluxes over a decently long period. Heat is transferred by a working fluid to another region of the heat pipe where the heat is radiated away. As with hot structures, the structures operate at high temperature. A principle sketch is shown in Fig. 1.5.

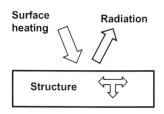

Figure 1.4 A hot structure.

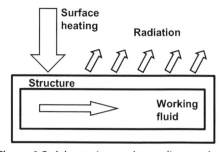

Figure 1.5 A heat pipe as the cooling method.

Ablation is another semipassive approach to handle thermal management. The purpose of the ablator is to keep the structure cool. Ablators are used for very high heat fluxes but for relatively short periods and are for single use. Fig. 1.6 shows a schematic of ablation cooling. Heat is also absorbed by the ablation process.

For still higher heat fluxes lasting over longer periods, active cooling is required. Convective cooling is often utilized in such cases. A schematic is shown in Fig. 1.7. Convective cooling is used in the propulsion system, in which heat is transferred to the coolant. The coolant is heated up and carries heat away. The structure operates at high temperature but the temperature is kept within limits by the active cooling process.

Film cooling, which is used inside a propulsion system, is another approach used for high heat fluxes lasting for long periods. The coolant is injected into the flow, commonly at an upstream location and at a single but discrete location. It operates as a thin, cool, and insulating blanket. The structure will maintain a high temperature (Fig. 1.8).

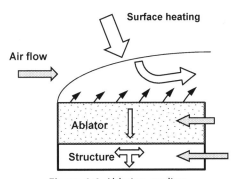

Figure 1.6 Ablation cooling.

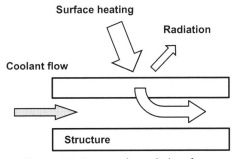

Figure 1.7 An actively cooled surface.

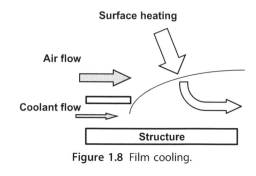

Figure 1.8 Film cooling.

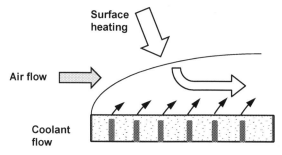

Figure 1.9 Transpiration cooling.

Transpiration cooling is another active cooling approach. This method is also used for high heat fluxes prevailing over long periods. Fig. 1.9 shows a schematic of transpiration cooling. The coolant is continuously injected into the hot gas flow through a porous structure over large areas as opposed to a discrete location in film cooling. The coolant decreases the heat flux to the structure but the structure will operate at a high temperature.

For air-breathing vehicles the TPS and the hot structure are highly integrated. However, there are differences between rocket-based systems and air-breathing vehicles. These differences impact the handling of thermal management and the TPS. Rockets accelerate but do not cruise while in atmospheric flight and are usually launched vertically. They leave the atmosphere quickly and generally fly at a low dynamic pressure. Conversely, air-breathing vehicles accelerate and cruise in the atmosphere. They are often launched horizontally and fly at a high dynamic pressure because they fly low in the atmosphere at high velocities to capture the air for the engine. For rocket-based vehicles, high drag is not a problem on ascent and is desired on descent deceleration. In contrast, air-breathing vehicles are optimized for low drag and thus have thin, slender bodies with low thickness-to-chord ratio. Rockets are extremely weight sensitive, whereas air-breathing vehicles tend to be more volume sensitive, which has an impact on the drag. Weight is also important for air-breathing vehicles. Rockets are driven by high-descent heating. On the other hand, air-breathing vehicles are driven by ascent, descent, and cruise, and they are exposed to high heat loads because of their long ascent times at high dynamic pressure. On a rocket the leading edge is usually blunt to facilitate a high descent drag and low heat flux. The air-breathing vehicles have sharp leading edges, as low drag and low thickness-to-chord ratio are desirable.

A thermostructural challenge is related to the occurrence of large thermal gradients. For instance, in a cryogenic tank containing liquid hydrogen as a fuel, the liquid hydrogen will be at a temperature about $-250°C$ and the outer surface of the thermal structure might be at a temperature between $1100°C$ and $1700°C$. As different materials have to operate at a wide range of temperatures, attaching various components (tank, insulation, structure, TPS, etc.), which expand and shrink, becomes a real challenge. In addition, production costs, life cycle cost, and inspection and maintenance costs need consideration.

The mentioned requirements have led to new approaches for the thermal protection, and development of new material systems is necessary to improvise the vehicles. For hypersonic air-breathing vehicles the materials must have high temperature capability, high strength at elevated temperatures, high toughness, lightweight, and environmental durability. A requirement for the materials is that they should maintain a high specific strength (strength divided by density) at elevated temperatures. Metallic based options include metallic matric composites (MMCs), super alloys, and titanium. At higher temperatures, ceramic matrix composites (CMCs), C-SiC material, advanced carbon-carbon (ACC), and SiC—SiC provide high strength.

The outer surface of hypersonic vehicles is subjected to severe aerodynamic heating. Accordingly, the airframe must be protected from the heat or designed to operate even when exposed to extreme heat. The components subjected to extreme heat are, e.g., the leading edges, TPS, aeroshells, and control surfaces. The leading edge radius has a significant impact on the heat flux. The heat flux is proportional to the reciprocal of the square root of the radius. This means that as the radius becomes smaller, the heat flux increases significantly. A sharp leading edge has an impact on the chordwise heat flux distribution; besides, the radius affects the maximum heat flux value. For sharp leading edges, passive, semipassive, and active thermal management options may be utilized to manage the intensive local heating.

High-temperature coatings based on SiC can be used up to about 1700°C. For higher temperatures, different materials such as carbides, oxides, and diborides of Hf and Zr can be used. The latter ones can be used as part of a composite matrix. Iridium, Ir, is regarded appropriate for coating. The thermal conductivity of the material may also affect the surface temperature, as carbon fiber is used and woven in various architectures.

Surface emissivity is another important parameter affecting the surface temperature. In free space, absence of conduction and convection represents a unique heat transfer problem. In high-vacuum conditions, the molecules are few and far apart and can only transport a negligible amount of energy. Therefore, thermal radiation is the only significant mode of heat exchange. A space vehicle can exchange radiant energy with the sun, with a nearby planet such as the Earth, and with the vast expanse of outer space. As a simple example to illustrate spacecraft temperature control, a spherical space vehicle is considered. It is assumed to have a uniform temperature and operate far from any planet. Radiation from the sun, G_s, is striking the space vehicle from one side. The amount of energy absorbed by the sphere is equal to the product of the solar radiation; the projected area of the sphere, $\pi D^2/4$; and the absorptance of the sphere for solar radiation, α_s,

$$\dot{Q}_{abs} = G_s \frac{\pi D^2}{4} \alpha_s \tag{1.5}$$

The radiation from the remainder of the space seen by the vehicle is assumed to be negligible because of the assumption that no planets are nearby. Outer space has an apparent temperature very near to absolute zero and thus emits very little energy. It also acts like a blackbody, reflecting none of the sphere's emitted energy. The energy emitted by the sphere is given by

$$\dot{Q}_{emitted} = \varepsilon \sigma \pi D^2 T^4 \tag{1.6}$$

where ε is the total emittance of the sphere surface at the temperature T. The sphere is radiating energy in all directions, whereas it is receiving energy only on one side from the sun. This will cause the sphere to have a hot and a cold side similar to the moon. Rotation of the sphere at a sufficient rate could help to maintain the isothermal temperature, which was initially assumed.

As $\dot{Q}_{abs}$ must be equal to $\dot{Q}_{emitted}$, one finds for the temperature of the spherical spacecraft

$$T = \left[\frac{\alpha_s}{\varepsilon} \frac{G_s}{4\sigma} \right]^{\frac{1}{4}} \tag{1.7}$$

The spacecraft's temperature depends on the magnitude of the solar radiation G_s and the α_s/ε ratio. The solar radiation depends on the distance from the sun and decreases with the square of distance from center of the sun. The ratio of the spacecraft surface absorptance to sunlight to its emittance (α_s/ε) determines the temperature at any fixed distance from the sun. By special paints and surface treatment, the value of α_s/ε can be controlled, and accordingly, the spacecraft temperature can also be controlled.

Recombination is an important process. In hypersonic flow, there are disassociated species from, e.g., air and recombination will occur. The surface will then serve as a catalyst to increase the recombination. This recombination can be exothermic, which means the reaction is releasing heat. Heat will be released at the surface and thus heat is provided to the surface. In addition to recombination of the gas flow, recombination of surface atoms can result in an increase in the heat flux. Silicon gas released from the surface coating reacting with an oxygen atom to form SiO is an example of an exothermic reaction, and this reaction releases a certain amount of heat. Commonly a recombination efficiency is introduced, and if it is close to zero, the surface is said to be noncatalytic and the heat flux is low. For a fully catalytic surface, the recombination efficiency is unity and the heat flux is much higher.

Thermal stresses are generated by the thermal expansion of a material, a temperature differential, and a structural or mechanical restraint of the thermal growth. This may also have an important impact on the design. Oxidation is another key factor in high-temperature structures because of the requirement of long lifetimes. To maximize the lifetime of the structure, it should be operated in the passive oxidation regime. Silicon carbide is a common coating material on many refractory composite structures. If there is a high oxygen pressure, then there is plenty of oxygen. The silicon carbide reacts with two oxygen molecules to form silicon dioxide (silica), which creates a protective barrier for the structure. Silica is a solid and forms a protective scale on the surface. CO_2 gas is emitted as a by-product. If the operating pressure is low, active oxidation can occur at high temperatures. In a low-pressure environment, there is not enough oxygen available. The result is that silicon oxide is formed instead of silica. This leads to active oxidation as silicon oxide is a volatile gas and leaves the surface without creating a protective scale.

Structurally integrated protection systems are thermally integrated and have higher structural efficiency, and thus there is a potential for lower maintenance. A principle sketch is shown in Fig. 1.10. The outer and inner

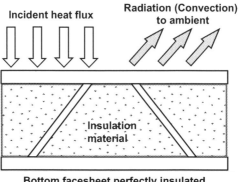

Figure 1.10 A structurally integrated TPS.

walls carry the airframe loads, with the outer wall operating hot, whereas the inner wall is insulated. The outer surface is made of a high-temperature material. The wall thickness provides stiffness, and the large integrated structure eliminates or reduces surface gaps and steps.

Control surfaces are required to control the vehicle during flight. Because drag is very important, attempts have been made to make the control surfaces thin. Different approaches for design and fabrication prevail, so insulated surfaces, hot structures, and hybrid solutions exist.

1.2.2 Cryogenic Matters

Cryogenic engineering concerns the production of very low temperatures and the effects created by low temperatures. Cryogenic temperature is commonly taken to be any temperature below $-150°C$. Substances used to obtain or maintain such temperatures are referred to as cryogens, which are normally liquefied gases. Common cryogens are liquefied nitrogen, hydrogen, oxygen, and helium. Cryogens have become particularly important because of their use in space propulsion systems, in medicine and surgery, and in studies of basic physical phenomena at low temperatures. The storage of liquefied gases is commercially important, as the liquid phase occupies much less volume per kilogram than the gaseous phase. Many cryogens can be treated as ordinary liquids and many common equations are applicable. A problem in dealing with solid substances at cryogenic temperatures is the large variations of the thermophysical properties. The large variations complicate analytical solutions of heat transfer problems; hence, numeric methods need to be utilized. In convective heat transfer problems with cryogens, a major difficulty appears because the critical point of the fluid may be approached or exceeded in many cases. This is a region with large variations in properties and this makes the analysis of flow and heat transfer difficult. The cryogens are always in a hostile environment because by definition they are colder than their surroundings. Boiling occurs readily because of this large temperature difference. Film boiling is a common occurrence, particularly in the cooldown of equipment. Two-phase flow and the accompanied unsteadiness can be expected in many situations. The boiling correlations for noncryogens are not suitable, and instead, experiments with cryogens are needed.

The only cryogen behaving in an unusual manner in terms of fluid flow and heat transfer is helium II. Below a temperature of 2.2 K, referred to as the lambda point, the fluid has very unusual characteristics such as superconductivity and superfluidity, i.e., extremely high thermal conductivity and low viscosity.

The storage of liquefied gases at low temperatures demands insulations with much lower thermal conductance than ordinary insulations. This has led to the development of new insulation materials and techniques. A significant result is the development of the so-called superinsulation materials. They consist of many layers of highly reflective material separated by low-conductivity spacers of some type and evacuated to low pressures. Further details can be found in Ref. [6].

1.2.3 Low-Density Heat Transfer

In continuum flow, the mean free path length of the molecular motion is extremely small compared to the boundary layer thickness, and analysis at the macroscopic level with the energy and the Navier–Stokes equations yields valid solutions. Under continuum conditions, flow and heat transfer phenomena can adequately be described in terms of dimensionless numbers such as the Reynolds, Mach, Nusselt, and Prandtl numbers. However, at small pressures and with increasing gas rarefaction, the gas partly loses its continuum characteristics and the intermolecular collisions become lesser, and thus the macroscopic analysis is not feasible. Accordingly, the molecular structure of the gas needs to be taken into account. In the free-molecular regime, the frequency of intermolecular collisions becomes much less and analysis is possible at microscopic level by using the kinetic gas theory with important parameters such as the mean free molecular path and the mean molecular velocity. Thus the basic formulation of the flow phenomena and heat transfer in the continuum and free-molecular regimes is much more refined than that of the other regimes of rarefied flows, such as the transitional and slip flow regimes. See also Ref. [7].

1.2.4 Gravity Effects

The topic of microgravity is highly relevant to aerospace applications. To the public, it is known as a weightless state. However, heat transfer under microgravity differs significantly from cases where normal gravity is present. Heat conduction in solids and liquids is not affected by gravity but heat conduction in gases is indirectly reduced in low gravity, because the gas density is reduced. Thermal radiation is not affected by gravity. However, phase change processes such as evaporation, boiling, and condensation, as well as two-phase forced convection and phase change heat transfer are affected by gravity. As the buoyancy forces become insignificant, other matters become important, such as capillary forces, viscous forces, and electromagnetic forces. Solidification under microgravity is of interest in materials science. Another important topic in aerospace application is the thermal stratification and pressurization in cryogenic tanks partially filled with LH_2 or LO_2 under reduced gravity. See also Ref. [8].

1.2.5 Heat Pipes

Heat pipes are attractive components in spacecraft cooling and temperature stabilization because of their low weight, limited maintenance request, and reliability. The subject of heat pipes has received great interest among researchers and engineers for use in a variety of applications. Heat pipes involve a variety of complex physical phenomena and require the involvement of several disciplines, such as thermodynamics, heat transfer, fluid mechanics, and solid mechanics. Spacecraft thermal management is challenged by adverse environmental thermal radiation. Heat pipes have been developed for both cryogenic and high-temperature applications. Also loop heat pipes and capillary pumped loops with several evaporators and condensers have been found to be effective in thermal management of high-powered communication satellites. Micro and micro-loop heat pipes are conjectured to be important for thermal management of small satellites. Additional information can be found in Refs. [9,10].

1.2.6 Auxiliary Equipment

There are several auxiliary equipment in aerospace applications. Here, a brief introduction to heat exchanger technology and fuel cells will be presented. Later chapters will present more details.

1.2.6.1 Heat Exchangers

Thermal performance requirements for aerospace applications are quite challenging and demanding. To meet such requirements, development of novel heat transfer enhancement techniques and design concepts is needed, as weight and volume allotments remain constant or even shrink. Compact heat exchangers with low weight are needed to meet the high heat flux removal requirements. Innovative heat transfer enhancement techniques are considered for the development of thermal management components. However, there seems to be no universal technology for all requirements. Heat exchangers based on microchannels have been proposed, as these offer size and weight benefits at a given performance compared to plate-fin and offset plate-fin devices. Application of high-porosity open cell metallic, carbon, and graphite foams has also been suggested. To achieve the benefits of such techniques, advanced manufacturing methods are needed to enable microstructured flow paths, which in turn provide large surface area-to-volume ratios and boundary layer disruption. Recently heat recovery in aviation engines has gained interest, as innovative developments for reducing the aviation greenhouse gas emissions are attempted. Further details, see Refs. [11,12].

1.2.6.2 Fuel Cells

The first practical application of fuel cells in space was for the Gemini project in the 1960s. A polymer electrolyte fuel cell (PEFC) was used. The membrane for the fuel cell was polystyrene. Later on the PEFC was replaced by an alkaline fuel cell system, but recently the PEFC has again gathered attention because of its applicability to ground facilities. As a spacecraft is very isolated in the Earth's orbit, all reactant materials must be carried inside the spacecraft. To minimize the weight, pure anode and cathode materials must be used and should be consumed completely. The interest is then also to use pure hydrogen and oxygen as reactants, avoid humidification of the reactant gases before being supplied to the fuel cell, consume as much reactant as possible, and collect the produced water. Successful tests have shown that a humidifier is not needed. Peroxide-based fuel cells have also been considered. The peroxide H_2O_2 is used at the cathode side and hydrogen or $NaBH_4$ is used at the anode side. It has been found that the direct utilization of all liquid, H_2O_2, and $NabH_4$ at the electrodes results in a 30% higher voltage output than an ordinary H_2/O_2 fuel cell. It was then concluded that the peroxide-based fuel cell is highly suitable for space power application where air is not available and a high energy density fuel is essential. A proton-exchange membrane (Nafion) was used as the electrolyte. Further details can be found in Ref. [13,14].

1.2.7 Miscellaneous Topics and SBLI

The rapid developments in jet propulsion, gas turbines, and high-speed flight brought forward the importance of compressible flow. The principle difference between incompressible and compressible flows is that the density variations of the fluid need to be considered for compressible flow. The density varies significantly in compressible flow and this can result in the occurrence of strange phenomena, such as shock waves. A shock wave can be considered as a discontinuity in the properties of the flow field. The fluid crossing a shock wave, normal to the flow path, will experience a sudden increase in pressure, temperature, and density, accompanied by a sudden decrease in speed, from a supersonic to a subsonic range. The process is irreversible. If the shock wave is not perpendicular to the flow, an oblique shock, the flow direction will also be changed. The thickness of the shock wave is of the order of only a few mean free paths. For external aerodynamics, usually a thin boundary layer prevails along the object surface. Interaction of the shock wave and boundary layer is of great importance and a lot of research in this area [shock wave—boundary layer interaction (SBLI)] has been carried out. Further details, see Ref. [15].

REFERENCES

[1] Holman JP. Heat transfer. 10th ed. New York: McGraw-Hill; 2009.
[2] Incropera FP, DeWitt DP, Bergman T, Lavine A. Introduction to heat transfer. 5th ed. New York: J. Wiley & Sons; 2007.
[3] Sunden B. Introduction to heat transfer. Southampton (UK): WIT Press; 2012.
[4] Modest MF. Radiative heat transfer. 2nd ed. New York: Academic Press; 2003.
[5] Howell JR, Siegel R, Menguc MP. Thermal radiation. 5th ed. Boca Raton: CRC Press, Taylor & Francis; 2011.
[6] Flynn T. Cryogenic engineering. 2nd ed. CRC Press; 2004.
[7] Frohn A, Roth N, Anders K. M10 heat transfer and momentum flux in rarefied gases. VDI Heat Atlas 2010:1375—90.
[8] Chalandon X, Webbon B, Montgomery L. Heat transfer and mass transfer in microgravity and hypobaric environments. In: SAE Technical Paper-941318; 1994. 17pp. http://dx.doi.org/10.4271/941318.
[9] Faghri A. Heat pipes: review, opportunities and challenges. Front Heat Pipes 2014;5(1):1—48.
[10] Shukla KN. Heat pipes for aerospace applications — an overview. J Electron Cool Therm Control 2015;5:1—14.
[11] Williams M, Muley A, Bolla J, Strumpf H. Advanced heat exchanger technology for aerospace applications, paper SAE 2008-01-2903. 2008.

[12] Heltzel AJ. In: Simulation of emerging heat exchanger technologies for progressive aerospace platforms, AIAA2010−291, 48th AIAA aerospace sciences meeting, January 2010, Orlando, Florida; 2010.

[13] Sone Y, Ueno M, Kuwajima S. Fuel cell development for space applications: fuel cell system in a closed environment. J Power Sources 2004;137:269−76.

[14] Luo N, Miley GH, Mather J, Burton R, Hawkins G, Gimlin R, Rusek J, Valdez TI, Narayanan SR. $NaBH_4/H_2O_2$ fuel cells for Lunar and Mars exploration. AIP Conf Proc 2006:209−21.

[15] Bhagwandin V, DeSpirito J. In: Numerical prediction of supersonic shock-boundary layer interaction, AIAA2011−859, 49th AIAA aerospace sciences meeting, January, 2011, Orlando, Florida; 2011.

CHAPTER 2

Ablation

2.1 INTRODUCTION

In aerospace design, ablation is used to both cool and protect mechanical parts that would otherwise be damaged by extremely high temperatures. Examples are heat shields for spacecraft, satellites, and missiles entering a planetary atmosphere from space and cooling of rocket engine nozzles. Basically, ablative material is designed to slowly burn away in a controlled manner, so that heat can be removed from the spacecraft by the gases generated by ablation, whereas the remaining solid material insulates the spacecraft from superheated gases. There has been an entire branch of spaceflight research involving the search for new fireproof materials to achieve the best ablative performance. Such a function is critical to protect the spacecraft occupants from otherwise excessive heat loading. The physical phenomena associated with ablation heat transfer depend on the application, but most involve in-depth material pyrolysis (charring) and thermochemical surface ablation. In numerical modeling, it is required to solve an energy equation, including effects of pyrolysis on a domain that changes as the surface ablates. Fig. 2.1 gives a general view of the ablation

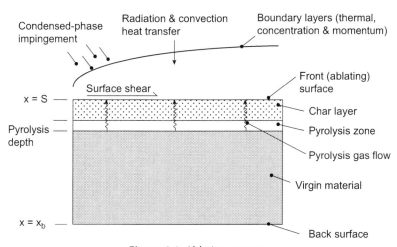

Figure 2.1 Ablation process.

Heat Transfer in Aerospace Applications
ISBN 978-0-12-809760-1
http://dx.doi.org/10.1016/B978-0-12-809760-1.00002-8

process and the involved mechanisms. As the material is heated, one or more components of the original composite material pyrolyze and yield a pyrolysis gas and a porous residue. The pyrolysis gas percolates away from the pyrolysis zone. The residue is often a carbonaceous char. The solution procedure is in principle a transient heat conduction calculation coupled to a pyrolysis rate calculation and subject to boundary conditions from the flow field.

In terms of heat transfer, the following mechanisms are involved: (a) convection in the boundary layer, which gives rise to the main thermal load; (b) radiation; and (c) conduction in the virgin material. In addition, resin decomposition and fiber decomposition may occur. In the boundary layer, shocks may appear, and in some cases, there may be combustion.

Ablation is affected by the freestream conditions, the geometry of the reentry body, and the surface material. The vehicles range from blunt configurations, such as spacecraft, to slender sphere-cone projectiles. At low heating values, ablators of Teflon are used, whereas at high heat loads, graphites and carbon-based materials are used. The most common ablative materials are composites, i.e., materials consisting of a high–melting–point matrix and an organic binder. The matrix might be glass, asbestos, carbon, or polymer fibers braided in various ways. A honeycomb structure filled with a mixture of organic and inorganic substances and with high insulating characteristics can be used. Advances in chemistry and materials technology extend the possibilities of selecting improved ablative materials.

Also surface movement occurs by, e.g., material spallation. Various methods exist to solve the moving boundary problems, commonly referred to as the Stefan problems. Basically two methods are considered: the front-tracking methods and the front-fixing methods. With the front-tracking methods, the ablating surface (the front) is tracked as it moves into the material.

2.2 AN ILLUSTRATIVE EXAMPLE OF ABLATION

In this section a simplified model of ablation is presented. Basically a transient heat conduction analysis is presented. The transient thermal response of the material exposed to a high-energy environment is important knowledge in the design of heat shields for reentry vehicles. The surface of a semi-infinite solid is heated by applying a constant heat flux q_0 (caused by frictional or aerodynamic heating) as shown in Fig. 2.2.

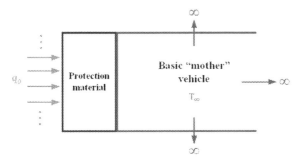

Figure 2.2 Simple illustration of an ablation process.

At time $\tau = 0$ the surface temperature has risen to the melting temperature T_p and phase change is initiated. The melted material (liquid) formed is completely removed by the aerodynamic forces. In this case the surface recedes with time but the surface temperature remains constant at the phase change temperature. A temperature distribution exists only in the remaining solid as conjectured in Fig. 2.3.

At a certain time the solid surface is located at $x = X(\tau)$. The temperature variation in the solid material penetrates to a depth $x = \delta(\tau)$. The temperature of the solid at far distances, $x > \delta(\tau)$, from the surface is kept at the constant initial temperature T_∞.

Introduce $\theta(x, \tau) = T - T_\infty$, which implies that $\theta(X, \tau) = T_p - T_\infty = \theta_p$. The one-dimensional unsteady heat conduction is then governed by

$$\frac{\partial \theta}{\partial \tau} = \alpha \frac{\partial^2 \theta}{\partial X^2} \qquad (2.1)$$

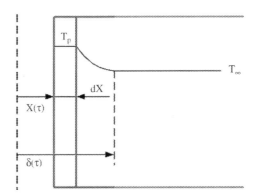

Figure 2.3 Temperature distribution in the simplified ablation process.

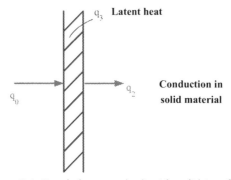

Figure 2.4 Heat balance at the liquid—solid interface.

In Fig. 2.4 the heat balance at the interface is presented.
The thermal balance at the interface can be expressed as below:

$$q_0 - q_2 = q_3 \tag{2.2a}$$

$$q_0 + k\frac{\partial\theta}{\partial X} = \rho Q_L\frac{dX}{d\tau} \tag{2.2b}$$

The solution of Eq. (2.1) can be found by some standard procedures but here the integrated form is used. Details are given below.

The integrated form of Eq. (2.1) can be written as

$$\frac{d}{d\tau}\left[\int_X^\delta \theta(x,\tau)dx - \theta(\delta,\tau)\cdot X\delta(\tau) + \theta(X,\tau)X(\tau)\right]$$

$$= \alpha\left[\frac{\partial\theta}{\partial x}(\delta,\tau) - \frac{\partial\theta}{\partial x}(X,\tau)\right] \tag{2.3}$$

A second-order polynomial temperature profile is assumed as

$$\frac{\theta(x,\tau)}{\theta_P} = \left[1 - 2\frac{x - X}{\delta - X} + \left(\frac{x - X}{(\delta - X)}\right)^2\right] \tag{2.4}$$

With the conditions $\theta(\delta,\tau) = 0$ and $\theta(X,\tau) = \theta_P$.

By combining the interface heat balance Eq. (2.2) with Eq. (2.3) and the assumed temperature profile Eq. (2.4), one obtains

$$\frac{d}{d\tau}\left[\frac{1}{3}(\delta - X)\right] + \left(1 + \frac{Q_L}{c\theta_P}\right)\frac{dX}{d\tau} = \frac{q_0}{\rho c\theta_P} \tag{2.5}$$

Further combination of the equations gives

$$\rho QL \frac{dX}{d\tau} = q_0 \frac{2k\theta_P}{-\delta - X} \tag{2.6}$$

Eqs. (2.5) and (2.6) are simultaneous equations, making the solutions of $X(\tau)$ and $\delta - X$ possible. The initial condition for X is $X(0) = 0$. The initial condition for $\delta(\tau)$, the specification for δ when the surface reaches the melting temperature T_P, can be obtained from solutions of a semi-infinite solid exposed to a time varying surface heat flux by the equations (see, e.g., Ref. [1])

$$\delta(\tau) = \sqrt{6\alpha\tau}$$

which gives the value of $\delta(\tau)$ if a parabolic temperature profile is assumed, and

$$\delta(\tau) = \sqrt{1.5\alpha}$$

which gives the specification for the surface temperature as a result of a constant heat flux input q_0.

Using the above equations the value of $\delta(\tau_P)$ when the surface temperature reaches θ_P may be calculated as

$$\delta(\tau_P) = 2\frac{k\theta_P}{q_0} \tag{2.7}$$

which is the remaining required initial condition.

Eqs. (2.5) and (2.6) have a steady-state solution if $dX/d\tau = A$ is a constant. Thus from Eq. (2.6), the value of $\delta - X$ must be constant, and then Eq. (2.5) gives

$$\frac{dX}{d\tau} = A = \frac{q_0}{\rho(Q_L + c\theta_P)} \tag{2.8}$$

Goodman [2] has solved Eqs. (2.5) and (2.6) by eliminating $dX/d\tau$ between them to obtain finally

$$\Omega = -\frac{1}{3}\left[\zeta - 2 + 2(1 + v)\ln\frac{2(1 + v) - \zeta}{2v}\right] \tag{2.9}$$

where

$$\Omega = \frac{\tau q_0^2}{\rho k\theta_P Q_L}, \quad v = \frac{Q_L}{c\theta_P}, \quad \zeta = \frac{(\delta - X)q_0}{k\theta_P}$$

Substituting Eq. (2.9) into Eq. (2.6) yields

$$\lambda = -\frac{1}{3}\left[\zeta - 2 + 2v\ln\frac{2(1+v)-\zeta}{2v}\right] \qquad (2.10)$$

where $\lambda = \frac{Xq_0}{k\theta_p}$.

The results above are depicted in Fig. 2.5. It presents, in dimensionless form, the melt-line location versus time, and the parametric influence of the variable v introduced as $v = Q_L/(c \cdot \theta_p)$ is also clearly depicted.

2.3 ADDITIONAL INFORMATION

Other simplified analyses have been presented by Han [3] and Holman [4]. In the analyses, it was assumed that a steady-state situation is attained and that the surface ablates at a constant rate.

Problems related to a melting solid for a variety of other boundary conditions have been presented in the literature. For instance, vaporization at the surface and aerodynamic heating by thermal radiation have been considered. Murray and Russell [5] presented a method for coupled aeroheating and ablation for missile configurations. Surface temperature and transient heat conduction calculations were carried out. The heat transfer coefficient was assumed to follow a known time function. The calculation method was verified against the flight test data.

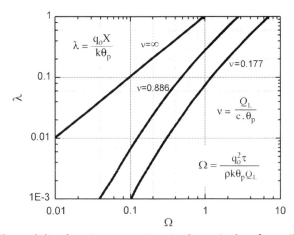

Figure 2.5 The melt-line location versus time in dimensionless form. *(Redrawn from Goodman Goodman R, The heat balance integral and its application to problems involving a change of phase. Trans. ASME 1958;80(2):335–345.)*

In Ref. [6], a method of thermal protection for transatmospheric vehicles exposed to high heating rates was introduced. The method involved combined radiation, ablation, and transpiration cooling. By placing an ablative material behind an outer shield that is of fixed shape and is porous, the effectiveness of the transpiration cooling was maintained, while retaining the simplicity of a passive mechanism. In the analysis, a simplified one-dimensional approach was used to derive the governing equations, which were reduced to a nondimensional form. In doing so, two parameters appeared to control the thermal protection effectiveness and ablation. The parameters are related to the thermal properties of the ablative and shield materials. The ablative material was also required to absorb a sufficient amount of thermal energy (related to the heat of ablation or vaporization) to keep the outer shield temperature below a specified value. A low vaporization temperature was found favorable to allow for the release of gaseous products and take advantage of the transpiration mechanism completely. Four ablative materials were considered and it was found that one material with thermal properties similar to those of Teflon behaved very well.

A relatively old report [7] presented a qualitative review of fundamental relationships involved in engineering and aerodynamic heating. The report included aerodynamic heating, boundary layer mass transfer, general thermal protection, ablative materials, and the structural aspects of ablation. In particular the general properties of charring ablators were discussed.

A comprehensive study of thermal protection systems was presented in a doctoral thesis [8], in which most mechanisms of the ablation phenomenon were treated in detail but mainly by computational fluid dynamics (CFD)-based methods.

A simplified analysis, similar to the one presented in this chapter, of ablation in cylindrical bodies was presented in Ref. [9]. The so-called general integral transform technique (GITT) was used to analyze the unsteady heat conduction, but with a transient surface heat flux.

REFERENCES

[1] Eckert ERG, Drake Jr RM. Analysis of heat and mass transfer. McGraw-Hill, Kogakusha; 1972.
[2] Goodman TR. The heat balance integral and its application to problems involving a change of phase. Trans ASME 1958;80(2):335—45.
[3] Han JC. Analytical heat transfer. New York: CRC Press; 2012.
[4] Holman JP. Heat transfer. 10th ed. New York: McGraw-Hill; 2009.

[5] Murray A, Russell G. Coupled aeroheating/ablation analysis for missile Neumann configurations. J Spacecr Rockets 2002;39(34):501−8.

[6] Camberos JA, Roberts L. Analysis of internal ablation for the thermal control of aerospace vehicles, Joint Institute for Aeronautics and Acoustics, JIAA TR-94. Stanford University; 1989.

[7] Achard RT. Fundamental relationships for ablation and hyperthermal heat transfer, AADL-TR-66−25. 1966.

[8] Bianchi D. Modeling of ablation phenomena in space applications. PhD thesis. Universita degli Studi di Roma "La Sapienza"; 2007.

[9] Aparecido Gomes FA, Campos Silva JB, Diniz JA. Heat transfer with ablation in cylindrical bodies. In: 24th International Congress of the Aeronautical Sciences. ICAS; 2004.

CHAPTER 3

Aerodynamic Heating: Heat Transfer at High Speeds

3.1 INTRODUCTION

In many engineering applications, it may be assumed that the influence on the energy balance by the work of the viscous forces is negligible. As the flow velocity becomes sufficiently high, this assumption will not be valid but instead the so-called frictional or viscous heating needs to be considered. Aerodynamic heating increases with the speed of the vehicle and it produces much less heat at subsonic speeds but becomes more significant at supersonic speeds. Aerodynamic heating is a concern for supersonic and hypersonic aircraft, as well as for reentry vehicles. The heating induced by the very high speeds greatly affects the overall design of aerospace vehicles. In addition the temperature gradients then become so large that the temperature dependence of the thermophysical properties also has to be considered. A thermal balance of aerodynamic heat input and thermal radiation heat loss to the sky for a flight at Ma = 4 at stratospheric altitudes results in an equilibrium surface temperature of about 800 C, which can be withstood by some steel materials. On the other hand, at hypersonic Mach numbers, there may be localized temperatures exceeding the melting temperatures of all substances. It can be estimated that for reentry vehicles the air temperature in the boundary layer may reach values above 5000 C. For hypersonic vehicles, it is generally known that aerodynamic heating is the dominant issue of the vehicle design. The heating effect is greatest at the leading edges and is dealt with by using high-temperature metal alloys, insulating the exterior of the vehicle, or using ablative materials.

The phenomenon of aerodynamic heating is, as stated earlier, of relevance and importance for aircraft, space vehicles, and missiles operating at supersonic velocity. Analysis of this type of problem is complicated. The purpose of this chapter is to introduce the phenomenon and present solutions for an idealized case. In many engineering cases involving high flow velocities along a flat plate and constant wall temperature, the wall heat flux can be calculated by the heat transfer coefficient, h, for the low-velocity

Heat Transfer in Aerospace Applications
ISBN 978-0-12-809760-1
http://dx.doi.org/10.1016/B978-0-12-809760-1.00003-X

case multiplied by the temperature difference, $T_w - T_{aw}$, where T_w is the wall temperature and T_{aw} is the so-called adiabatic wall temperature. The theory and background will be summarized. Examples are provided to illustrate the phenomenon.

3.2 HIGH VELOCITY FLOW ALONG A FLAT PLATE

Consider the high-speed motion of an incompressible fluid along a flat plate as depicted in Fig. 3.1.

The fluid is assumed to have constant thermophysical properties. According to Appendix 1, the laminar boundary layer equations read as follows:

Mass

$$\frac{\partial u}{\partial x} + \frac{\partial v}{\partial y} = 0 \tag{3.1}$$

Momentum

$$u\frac{\partial u}{\partial x} + v\frac{\partial u}{\partial y} = v\frac{\partial^2 u}{\partial y^2} \tag{3.2}$$

Energy

$$u\frac{\partial T}{\partial x} + v\frac{\partial T}{\partial y} = \alpha\frac{\partial^2 T}{\partial y^2} + \frac{\mu}{\rho\, c_p}\left(\frac{\partial u}{\partial y}\right)^2 \tag{3.3}$$

The last term in Eq. (3.3) represents the work by the viscous forces or the so-called frictional or viscous heating.

The boundary conditions of Eqs. (3.1)−(3.3) read

$$y = 0: u = v = 0, \quad T = T_w \tag{3.4}$$

$$y \to \infty : u \to U_\infty, \quad T \to T_\infty \tag{3.5}$$

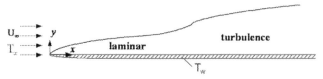

Figure 3.1 Boundary layer flow along a flat plate surface.

The flow field is not changed when compared to the low-speed case. The solution can be found in most heat transfer text books like Refs. [1,2].

The solution of the temperature field is split into two: a homogeneous solution and a particular solution. The homogeneous solution is obtained when the frictional heating term in Eq. (3.3) is neglected. The homogeneous solution of Eq. (3.3) can be written as

$$T^{(h)} = T_\infty + (T_w - T_\infty)\theta(\eta) \tag{3.6}$$

where the dimensionless coordinate η is given by, $\eta = \sqrt{\frac{U_\infty}{2\nu x}}\, y$.

As a particular solution of Eq. (3.3), one looks for the solution being valid for an adiabatic plate, i.e., $q_w = 0$. With η defined as earlier and the approach of the concept of the stream function ψ ($u = \partial\psi/\partial y$, $v = -\partial\psi/\partial x$, $f' = u/U_\infty$) is represented as

$$\psi = \sqrt{2\nu U_\infty x} f(\eta)$$

Eq. (3.3) can then be written as

$$\frac{d^2 T}{d\eta^2} + Prf\frac{dT}{d\eta} + 2\frac{U_\infty^2}{2c_p}Prf''^2 = 0 \tag{3.7}$$

The boundary conditions of Eq. (3.7) are

$$\eta = 0: \frac{dT}{d\eta} = 0 \quad (q_w = 0)$$

$$\eta \to \infty : T \to T_\infty$$

A dimensionless temperature θ^a is defined in Eq. (3.8)

$$\theta^a = \frac{T - T_\infty}{U_\infty^2/2c_p} \tag{3.8}$$

By introducing θ^a in Eq. (3.7), one obtains

$$\frac{d^2\theta^a}{d\eta^2} + Prf\frac{d\theta^a}{d\eta} + 2Pr\, f''^2 = 0 \tag{3.9}$$

The boundary conditions are transferred as

$$\eta = 0: \frac{d\theta^a}{d\eta} = 0 \tag{3.10}$$

$$\eta \to \infty : \theta^a \to 0 \tag{3.11}$$

Eq. (3.9) with the boundary conditions in Eqs (3.10) and (3.11) was solved by Eckert [5] many years ago. It was shown that the difference between the adiabatic wall temperature (the plate temperature for $q_w = 0$) and the fluid temperature could be written as

$$T_{aw} - T_\infty = r \cdot \frac{U_\infty^2}{2c_p} = \theta^a(0)\frac{U_\infty^2}{2c_p} \qquad (3.12)$$

where r is the so-called recovery factor, which depends on the Pr number of the fluid.

For moderate Pr numbers (e.g., gases, water) the recovery factor becomes

$$r \cong \sqrt{Pr} \quad \text{for} \quad 0.6 < Pr < 15 \qquad (3.13)$$

The solution of Eq. (3.3) with the conditions Eqs. (3.4) and (3.5) is obtained by combining the homogeneous and the particular solutions, i.e.,

$$T = T^{(h)} + T^{(p)}$$

or

$$T = C_1 T_\infty + C_2(T_w - T_\infty)\theta(\eta) + \theta^a\frac{U_\infty^2}{2c_p} + T_\infty$$

The boundary condition Eq. (3.5) requires $C_1 = 0$, whereas the boundary condition Eq. (3.4) implies

$$T_w = C_2(T_w - T_\infty) \cdot 1 + (T_{aw} - T_\infty) + T_\infty$$

The constant C_2 is found from this expression as

$$C_2 = \frac{T_w - T_{aw}}{T_w - T_\infty}$$

The solution of Eq. (3.3) can therefore be written as

$$T - T_\infty = (T_w - T_{aw})\theta(\eta) + \theta^a\frac{U_\infty^2}{2c_p} \qquad (3.14)$$

3.3 CALCULATION OF THE HEAT TRANSFER

The heat flux from the plate surface q_w is written as usual, i.e.,

$$q_w = -k\left(\frac{\partial T}{\partial y}\right)_w = -k\left(\frac{\partial T}{\partial \eta}\right)_w\sqrt{\frac{U_\infty}{2vx}}$$

From Eq. (3.14), one finds

$$\left(\frac{\partial T}{\partial \eta}\right)_w = (T_w - T_{aw})\left(\frac{\partial \theta}{\partial \eta}\right)_w + \left(\frac{\partial \theta^a}{\partial \eta}\right)_w \frac{U_\infty^2}{2c_p}$$

However, $\left(\frac{\partial \theta^a}{\partial \eta}\right)_w = 0$ and $\left(\frac{\partial \theta}{\partial \eta}\right)_w$ are the temperature derivatives for the low-speed case. From Ref. [1], the solution for the low-speed case can be found as

$$\left(\frac{\partial \theta}{\partial \eta}\right)_w = -\sqrt{2}\,\frac{h}{k}\sqrt{\frac{vx}{U_\infty}}$$

The heat flux can now be written as

$$q_w = h(T_w - T_{aw}) \tag{3.15}$$

Thus a simple method to calculate the heat flux at high flow velocities has been found.

The heat transfer coefficient for the corresponding low-speed case is multiplied by the temperature difference, $T_w - T_{aw}$, where T_{aw} is the so-called adiabatic wall temperature. This temperature is determined using the recovery factor r (Eq. (3.12)).

3.4 TURBULENT FLOW

The described analysis is valid for laminar boundary layers. For a turbulent boundary layer, it has been found that the heat flux at high velocities can be calculated similarly. Turbulent flow prevails where $Re_x = U_\infty\, x/v > 5 \cdot 10^5$. The recovery factor is determined by Ref. [4]

$$r = Pr^{1/3} \tag{3.16}$$

The heat transfer coefficient for turbulent low flow velocities can be found in most heat transfer text books, e.g., Refs. [1,2].

3.5 INFLUENCE OF THE TEMPERATURE DEPENDENCE OF THE THERMOPHYSICAL PROPERTIES

In Section 3.1, it was mentioned that at high flow velocities the thermophysical properties will be affected because of the large temperature gradients. This means that these properties may vary considerably across the boundary layer. A complete analysis is too extensive but Pohlhausen [3] recommends that the thermophysical properties are evaluated at a reference temperature T^* according to

$$T^* = T_\infty + 0.5(T_w - T_\infty) + 0.22(T_{aw} - T_\infty) \qquad (3.17)$$

3.6 TEMPERATURE DISTRIBUTION IN THE BOUNDARY LAYER

Fig. 3.2 presents a temperature profile in the thermal boundary on a flat surface. If the wall temperature is higher than the adiabatic wall temperature, heat is transferred from the surface to the gas. On the other hand, if the wall temperature is below the adiabatic wall temperature, heat is transferred from the fluid to the surface.

3.7 ILLUSTRATIVE EXAMPLE

Consider a flat plate, as in Fig. 3.1, with the dimensions, length, 70 cm and width, 1 m. The plate is placed in a wind tunnel where the flow conditions are given by Ma = 3, p = 0.05 atm, temperature $-40°$C. Two tasks are described in the following: (1) to calculate the recovery temperature and (2) to calculate the cooling rate to maintain the plate surface temperature at $35°$C.

1. The freestream sound velocity, a, is first calculated as

$$a = \sqrt{\gamma RT} = (1.4 \cdot 287 \cdot 233)^{1/2} = 306 \ \text{m/s}$$

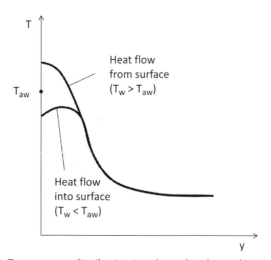

Figure 3.2 Temperature distribution in a boundary layer along a flat plate.

Then the freestream velocity is

$$U_\infty = 3 \cdot 306 = 918 \ \text{m/s}$$

The maximum Reynolds number is estimated using the properties at the freestream conditions:

$$\rho_\infty = p/RT = 0.0758 \ \text{kg/m}^3, \ \mu_\infty = 1.434 \cdot 10^{-5} \ \text{kg/ms};$$
$$Re_L = \rho_\infty U_\infty \ L/\mu_\infty = 3.395 \cdot 10^6$$

Thus one needs to consider both the laminar and the turbulent boundary layers.

Laminar boundary layer. With $Pr = 0.7$ the recovery factor becomes $r = \sqrt{Pr} = 0.837$. From Eq. (3.12), one obtains $T_{aw} = -40 + 0.837 \cdot 918^2/(2 \cdot 1005) = 311°C$.

Now the thermophysical properties should be recalculated based on the reference temperature in Eq. (3.17). However, it is found that the differences are small.

Turbulent boundary layer. According to Eq. (3.16) the recovery factor becomes $r = 0.7^{1/3} = 0.888$.

The corresponding adiabatic wall temperature is found to be $T_{aw} = -40 + 0.888 \cdot 918^2/(2 \cdot 1005) = 332°C$.

2. The cooling rate can be calculated as follows

$$Q_w = h \cdot A \cdot (T_{aw} - T_\infty)$$

The low-speed heat transfer coefficient needs to be found for the laminar and the turbulent boundary layers. By using the critical Reynolds number for transition as $5 \cdot 10^5$, it can be found that the length of the laminar boundary layer is 0.222 m and that of the turbulent boundary layer is 0.478 m.

Laminar boundary layer. The average heat transfer coefficient can be found from Ref. [1] $Nu = hL_{lam}/k = 0.664 \cdot \sqrt{Re} \cdot Pr^{1/3}$, which gives $h_{lam} = 56.3 \ \text{W/m}^2\text{K}$. Then the heat flow rate is $Q_{lam} = 56.3 \cdot (0.222 \cdot 1) \cdot (311 - 35) = 3445 \ \text{W}$.

Turbulent boundary layer. The corresponding heat transfer coefficient can be found from the local one given by Ref. [1], $Nu = h \cdot x/k = 0.0296 \ Re_x^{0.8} \cdot Pr^{1/3}$. However, the average value has to be found by averaging from $x = 0.222$ m to $x = 0.7$ m. The average value becomes $h_{turb} = 87.5 \ \text{W/m}^2\text{K}$. Then the heat flow rate is $Q_{turb} = 87.5 \cdot (0.47 \cdot 1) \cdot (332 - 35) = 12,415 \ \text{W}$.

The total amount of heat to be cooled from the plate becomes

$$Q_{tot} = Q_{lam} + Q_{turb} = 15860 \text{ W}.$$

3.8 AN ENGINEERING EXAMPLE OF A THERMAL PROTECTION SYSTEM

As mentioned in Section 3.1, space vehicles have to withstand extremely high aerodynamic heating and pressure loads during the ascent and reentry stages. The primary function of a thermal protection system (TPS) is to keep the underlying structure within acceptable temperature limits and to maintain the aerodynamic shape of the vehicle without excessive deformation (thermal bowing and bending caused by pressure loads). A TPS must provide thermal protection and also have structural load bearing capabilities and be robust, durable, and easily maintained. Although satisfying all these requirements, the weight of the TPS must be as low as possible to keep the launch costs low. A sandwich panel to be used as a TPS is shown in Fig. 3.3. Significant efforts have been taken toward the development of integrated thermal protection system (ITPS) that exhibits the function of thermal protection and load-bearing capabilities [6−10].

The objective of this example is to present findings from an investigation [11] that developed an analytical procedure to optimize and obtain a minimum-mass corrugated-core sandwich panel to function as an ITPS for space vehicles. The optimal dimensions for both thermal and structural constraints were attempted, but the two constraints were not active at the same instant of time. Accordingly the optimization problem was solved in two steps. In the first step, the design process for the corrugated-core ITPS was initiated by a transient thermal analysis to obtain the temperature distribution as a constraint and load. Seven dimensions of the unit cell were

Figure 3.3 A typical corrugated-core sandwich structure for thermal protection system (TPS), with top face sheet, web, and bottom faceplates.

considered as the geometric design variables. The thickness of the insulation was optimized in the thermal sizing study, and then the optimal height of the core was determined. In the second step, the geometry of the ITPS panel was optimized, except for the height of the core identified in the first step. The structural sizing optimization considered both thermal and mechanical loads. The globally convergent method of moving asymptotes (GCMMA) [12] was used as the optimization algorithm. The ANSYS Parametric Design Language (APDL) code was developed to analyze the thermomechanical coupling model.

3.8.1 Thermal Analysis

Consider a simplified geometry of an ITPS unit cell as shown in Fig. 3.4. The unit cell consists of two inclined webs and two thin faceplates. The top faceplate thickness, t_{TF}, may be different from the bottom faceplate thickness, t_{BF}, as well as the web thickness t_W. The unit cell is identified by seven geometric parameters (p, d, t_{TF}, t_{BF}, t_W, θ, and l). The thickness of the insulation is the only concern. The top and bottom faceplate thicknesses are constant because they have limited effect on the maximum temperature of the internal structure. The temperature on the top surface of the ITPS panel is always close to the radiation equilibrium temperature. It is controlled by the emissivity of the top surface, which typically has a value in the range of 0.8–0.85 [13]. Increasing the emissivity is not a design issue but is related to manufacturing and material selection. The thermal load and boundary conditions imposed on the two-dimensional (2D) model are shown in Fig. 3.5.

Furthermore, the following assumptions are made:

1. No temperature variation in the direction along the outer surface of the panel.

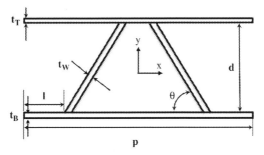

Figure 3.4 A simplified geometry of an ITPS unit cell: two inclined webs with the thickness t_W and two thin faceplates with the thickness t_{TF} and t_{BF}.

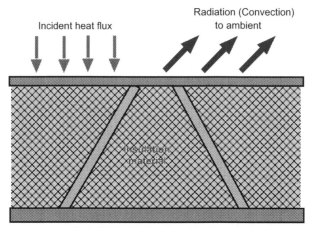

Incident heat flux

Radiation (Convection)
to ambient

Insulation
material

Perfectly insulated bottom face sheet

Figure 3.5 Loading and boundary conditions for a simplified ITPS structure model: instantaneous heat flux and radiation are assigned on the outer faceplate, whereas inner face sheet is perfectly insulated.

2. The lower surface of the bottom faceplate is assumed to be perfectly insulated. Optimization with this assumption leads to a conservative design.
3. Radiation is applied on the upper surface of the top faceplate, with an emissivity of 0.85.
4. Initial temperature of the model is assumed to be constant.
5. Radiative and convective heat transfer through the insulation material is ignored, and only the conductive heat transfer is taken into account.

3.8.2 Finite Element Analysis of Heat Transfer

The material thermal properties will change with temperature and pressure and accordingly the heat transfer problem turns into a nonlinear transient problem, which is relatively complicated. According to these assumptions, the governing heat conduction equation can be expressed as follows [1]:

$$\rho c \frac{\partial T}{\partial \tau} = \frac{\partial}{\partial x}\left(k\frac{\partial T}{\partial x}\right) \tag{3.18}$$

where ρ, k, and c is the density, thermal conductivity, and specific heat of the insulation material, respectively; τ is the time; and T is the temperature.

By discretization of Eq. (3.18), the finite element equation in the heat transfer analysis is transformed to the general form:

$$[\mathbf{C}(\mathbf{T})][\mathbf{T}'] + [\mathbf{K_c}(\mathbf{T})][\mathbf{T}] = [\mathbf{Q}(\mathbf{T})] \tag{3.19}$$

where $\mathbf{C}$ is the specific heat matrix; $\mathbf{K_c}$, conductivity matrix; $\mathbf{Q}$, nodal heat flow vector; and $\mathbf{T}$ and $\mathbf{T'}$, the nodal temperature vector and the derivative of the nodal temperature versus time, respectively. Each nodal transient temperature can be obtained using the Newmark method combined with the Newton–Raphson method to solve Eq. (3.19).

The thermal boundary conditions are as follows. The considered aerodynamic heat flux is under the assumption of laminar flow and is imposed on the upper surface of the top faceplate during reentry. The ambient temperature and initial structural temperature are both assumed to be 273 K. A large portion of heat is radiated out to the ambient by the upper surface of the top faceplate. The net thermal radiation is determined by

$$q_r = \varepsilon_s \sigma_s \left(T_t^4 - T_\infty^4 \right) \tag{3.20}$$

where q_r is the radiative heat flux; ε_s, the surface emissivity; σ_s, the Stefan–Boltzmann constant; T_t, the top faceplate temperature; and T_∞, ambient temperature.

After the vehicle has touched down, it is cooled only by natural convection. During this period, the temperature of the bottom faceplate rises to a maximum value. In this case, a convective heat transfer boundary condition is imposed on the upper surface of the top faceplate instead of radiation, as shown by Eq. (3.21):

$$q_c = h \left(T_t - T_\infty \right) \tag{3.21}$$

where q_c is the convective heat flux; h, the convective heat transfer coefficient; T_t, the top faceplate temperature; and T_∞, the ambient temperature. Here, the value of the convective coefficient, h, was assumed to be 6.5 W/m^2K [10].

In the transient thermal analysis, thermal properties, which may be a function of temperature and pressure, are updated at each time step. The aerodynamic pressure load on the vehicle during reentry is presented in Fig. 3.6 [6]. Based on the material properties database, the transient heat transfer analysis is divided into four load steps to fit the pressure profile.

The aerodynamic heating and natural convection coefficient profiles are presented in Fig. 3.7.

3.8.3 Thermal Results

In the studied case, a material combination based on Ref. [10] was selected. The materials from the top faceplate to the bottom faceplate are as follows: aluminosilicate/Nextel 720 composites (for top faceplate and web), Saffil

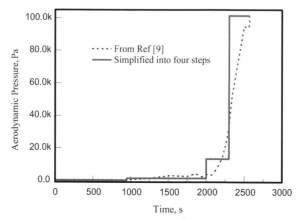

Figure 3.6 The aerodynamic pressure load on the vehicle during reentry: the instantaneous aerodynamic pressure is approximated by multiple-line variations. *(The pressure data was adapted from Grujicic M, Zhao CL, Biggers SB, Kennedy JM, Morgan DR. Heat transfer and effective thermal conductivity analyses in carbon-based foams for use in thermal protection systems. Proceedings of the Institution of Mechanical Engineers, Part L: Journal of materials: design and applications, 219(4):217–230; Martinez OA, Sankar BV, Haftka RT, Blosser ML. Two-dimensional orthotropic plate analysis for an integral thermal protection system, AIAA J 2012; vol. 50, pp. 387–398.)*

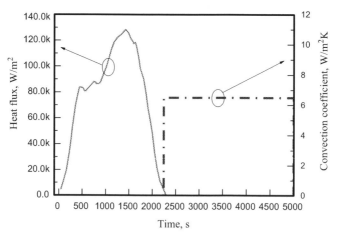

Figure 3.7 The aerodynamic heating and natural convection coefficient profiles: an instantaneous heat flux is imposed up to about 2300 s, whereas only constant natural convection is imposed beyond 2300 s.

insulation, and epoxy/carbon fiber laminate (for bottom faceplate). The material properties are taken from Refs [10] and [15] and are listed in Table 3.1.

Fig. 3.8 presents temperature variations versus reentry times for some featured points. It shows that the top faceplate reaches its peak temperature after 1500 s. When the aerodynamic convection-based input heat flux is balanced by the emitted radiative heat flux, the temperature of the top faceplate becomes equal to the radiation equilibrium temperature, which is determined by a given input heat flux and a given surface emissivity. The variations of the temperature along the thickness direction of the ITPS at several selected time steps are provided in Fig. 3.9. The temperature gradients of the structure at each characteristic time step are provided. At about 6350 s, the structure will not be heated any more, as the temperature gradient approaches zero.

A structural finite element model was used to calculate the stress of the top faceplate and the web. An assumed uniform pressure loading of 5000 Pa was applied on the sandwich outer surface, and temperatures are applied over the entire model, as illustrated in Fig. 3.10. The peak radiation equilibrium temperature for the reentry heat flux is 990 °C at 1500 s [15].

Table 3.1 Material Thermo Physical Properties

	Materials		
	Aluminosilicate-Nextel 720 Fiber Composite	**Epoxy/carbon Fiber Laminate**	**Saffil**
Properties			
Density, ρ (kg/m^3)	2450–2600	1550–1580	50
Young's modulus, E (GPa)	133.5–139.1	49.7–60.1	\
Compressive strength, σ_{bc} (MPa)	67.9–68.8	542.1–656.8	\
Tensile strength, σ_{bt} (MPa)	67.9–68.8	248.6–355.9	\
Poisson's ratio, μ	0.23–0.25	0.305–0.307	\
Specific heat, c (J/kgK)	950–1100	901.7–1037	942–1340
Thermal conductivity, k (W/mK)	2.52–2.93	1.28–2.6	0.052–0.5
Thermal expansion, α (10^{-6}/K)	5.745	0.36–4.02	\

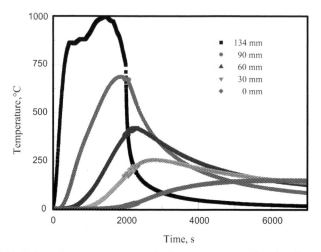

Figure 3.8 Variations in temperature with the reentry time for the featured points (distance from bottom: 0, 30, 60, 90, and 134 mm).

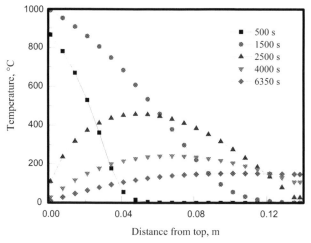

Figure 3.9 Temperature profiles in the thickness direction of the ITPS at several selected time steps: 500, 1500, 2500, 4000, and 6350 s.

Therefore the temperature distribution at 1500 s is chosen to simulate the situation where the thermal stress caused by the thermal gradient reaches its maximum value. In this calculation, the height of the sandwich core was fixed. The insulation material was not considered to be a structural member because the Saffil insulation is a soft fibrous insulation material with hardly any mechanical properties, but it was also modeled to simulate more realistic

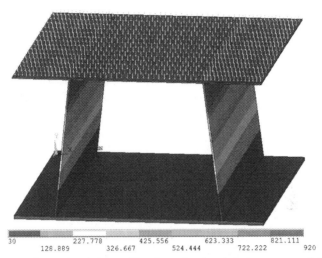

30   227.778   425.556   623.333   821.111
   128.889   326.667   524.444   722.222   920

Figure 3.10 Mechanical and thermal loads on the panel during reentry of the vehicle: a uniform pressure of 5000 Pa and temperature from the heat transfer analysis were applied on the top faceplate.

thermal expansion environment for the sandwich panel. In addition, temperature-dependent properties are ignored to further simplify the model and the material properties are considered to have fixed values.

The maximum von Mises stress of the ITPS unit cell occurs at four constraint points and is far less than the yielding strength of the epoxy/carbon fiber laminate. The maximum von Mises stress of the top faceplate and webs occurs at the edge of the webs.

Further details of this example of engineering relevance can be found in Ref. [16].

3.9 AERODYNAMIC HEAT REDUCTION

In Section 3.7 a TPS was analyzed. However, there are several studies concerning reduction of the aerodynamic heating by using active techniques such as an opposing jet, a forward-facing spike, or an opposing jet from an extended nozzle, as illustrated in Fig. 3.11. As the heat transfer mechanism in high enthalpy flow (supersonic and cold hypersonic flows) is different from that in low enthalpy flow, the design of a TPS is affected accordingly. As the flow speed exceeds the Mach number of unity, a shock wave is formed in front of the body and the flow temperature rises as it passes through it. As the temperature rises above a certain value, molecules in the air dissociate and

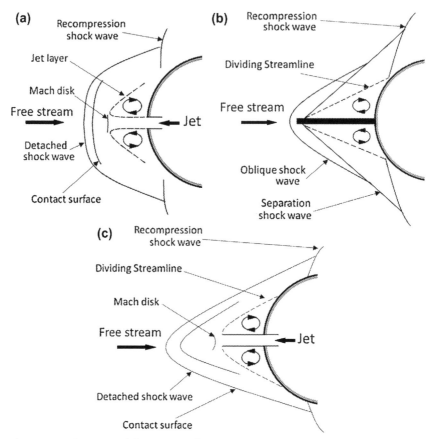

Figure 3.11 Conjectured flow pattern for active TPS by opposing jets. (a) opposing jet, (b) spike, and (c) opposing jet from extended nozzle.

become radicals. The radicals recombine on the body surface and then the surface is heated by radiation. The recombination rate on the surface depends on the catalysity of the surface. In turn the catalysity depends on the surface temperature and the substance of the surface. Accordingly the flow phenomena become complex by dissociation and recombination, but these processes reduce the flow temperature because of the loss of energy caused by dissociation. There is then a possibility that the aerodynamic heating is reduced by cutting down the convective heat transfer and the radiation heat transfer. It is known that oxygen and nitrogen molecules begin dissociation at a temperature of 2500 K and 4000 K, respectively. Thus for air, effects of dissociation and recombination have to be considered when the temperature backward of the shock reaches these temperatures.

Application of an opposing jet is a method to cool body surfaces in supersonic and hypersonic flows by injecting a coolant gas forward. Accordingly, many experimental investigations and CFD simulations have been applied in analyzing such systems to reduce the aerodynamic heating (as well as drag). It has been found that a strong opposing jet flow and a long extended nozzle reduce the aerodynamic heating and drag significantly. The opposing jet covers the body surface with the coolant gas and insulates the surface from the high-temperature gas, thus reducing the convective heat transfer. The radiative heat transfer is reduced by preventing recombination of the dissociated gas on the body surface. An additional cooling effect might be obtained close to the injection location because this region is protected by the jet and the recirculating flow. Relevant research works are described in Refs. [17−19].

REFERENCES

[1] Sunden B. Introduction to heat transfer. UK: WIT Press; 2012.
[2] Incropera FP, DeWitt DP, Bergman T, Lavine A. Introduction to heat transfer. 5th ed. New York: J. Wiley & Sons; 2007.
[3] Pohlhausen E. Der Wärmeaustausch zwischen festen Körpern und Flüssigkeiten mit kleiner Reibung und kleiner Wärmeleitung. Z Angew Math Mech 1921;1:115−21.
[4] Kaye J. Survey of friction coefficients, recovery factors and heat transfer coefficients for supersonic flow. J Appl Sci 1954;21:117−9.
[5] Eckert ERG. Engineering relations for heat transfer and friction in high-velocity laminar and turbulent boundary layer flow over surface with constant pressure and temperature. Trans ASME 1956;78:1273−84.
[6] Bapanapalli SK, Martinez OA, Gogu C, Sankar BV, Haftka RT, Blosser ML. Analysis and design of corrugated-core sandwich panels for thermal protection systems of space vehicles, 47th AIAA/ASME/ASCE/AHS/ASC Structures, Structural Dynamics, and Material Conference, Newport, Rhode Island, May 1−5, Paper No. 2006-1876.
[7] Martinez OA, Sankar BV, Haftka RT, Bapanapalli SK, Blosser ML. Micromechanical analysis of composite corrugated-core sandwich panels for integral thermal protection systems. AIAA J 2007;45:2323−36.
[8] Martinez OA, Sankar BV, Haftka RT, Blosser ML. Thermal force and moment determination of an integrated thermal protection system. AIAA J 2010;48:119−28.
[9] Martinez OA, Sankar BV, Haftka RT, Blosser ML. Two-dimensional orthotropic plate analysis for an integral thermal protection system. AIAA J 2012;50:387−98.
[10] Gogu C, Bapanapalli SK, Haftka RT, Sankar BV. Comparison of materials for an integrated thermal protection system for spacecraft re-entry. AIAA J Spacecr Rockets 2009;46:501−13.
[11] Poteet CC, Abu-Khajeel H, Hsu SY. Preliminary thermal-mechanical sizing of metallic TPS: process development and sensitivity studies. In: 40th Aerospace Sciences Meetings & Exhibitions, Reno, Nevada, January 14−17, Paper No. AIAA 2002−0505; 2002.
[12] Daryabeigi K. Heat transfer in high-temperature fibrous insulation. In: 8th AIAA/ASME Joint Thermophysics and Heat Transfer Conference, St. Louis, MO, June 24−26, Paper No. AIAA 2002−3332; 2002.

[13] Johnson TF, Waters WA, Singer TN, Haftka RT. Thermal-structural optimization of integrated cryogenic propellant tank concepts for a reusable launch vehicle. In: 45th AIAA/ASME/ASCE/AHS/ASC Structure, Structure Dynamics, and Materials Conference, Palm Springs, CA, April 19–22, Paper No. 2004–1931; 2004.

[14] Grujicic M, Zhao CL, Biggers SB, Kennedy JM, Morgan DR. Heat transfer and effective thermal conductivity analyses in carbon-based foams for use in thermal protection systems. Proc Inst Mech Eng Part L J Mater Des Appl, 219(4):217–230.

[15] Myers DE, Martin CJ, Blosser ML. Parametric weight comparison of advanced metallic, ceramic tile, and ceramic blanket thermal protection systems. NASA Technical Memorandum, Paper No. NASA TM-2000–210289.

[16] Xie G, Wang Q, Sunden B, Zhang W. Thermomechanical optimization of lightweight thermal protection system under aerodynamic heating. Appl Therm Eng 2014;59:425–34.

[17] Aruna S, Anjali Devi SP. A computational study on reduction of aerodynamic heating and drag over a blunt body in hypersonic turbulent flow using counter jet flow. Int J Aerodyn 2012;2(2/3/4):240–76.

[18] Imoto T, Okabe H, Aso S, Tani Y. Enhancement of aerodynamic heating reduction in high enthalpy flows with opposing jet. In: AIAA 2011-346 (49th AIAA Aerospace Sciences Meeting Including New Horizons Forum and aerospace Exposition, Orlando, Florida); 2011.

[19] Morimoto N, Aso S, Tani Y. Reduction of aerodynamic heating and drag with opposing jet through extended nozzle in high enthalpy flow. In: 29th Congress of the International Council of the Aeronautical Sciences (ICAS), St Petersburg, Russia; 2014.

FURTHER READING

Eckert ERG. Survey of boundary layer heat transfer at high velocities and high temperatures. WADC Tech Rep 1960:59–624.

Eckert ERG, Drake Jr RM. Analysis of heat and mass transfer. New York: McGraw-Hill; 1987.

CHAPTER 4

Low-Density Heat Transfer: Rarefied Gas Heat Transfer

4.1 INTRODUCTION

When an aerospace vehicle flies through the atmosphere at very high speeds, it is subjected to aerodynamic heating. The extent of heating depends, among other factors, on the density of air through which the vehicle is traveling. Aerospace missions involve design and deployment of vehicles that encounter all density flow regimes during their flight path. The physics of flight in the free molecular and continuum regimes has been reasonably well understood, whereas that in the transitional and slip flow regimes is still subjected to considerable study and research. In continuum flow, the mean free path length is extremely small compared to the boundary layer thickness and macroscopic analysis with the energy and Navier–Stokes equations yields valid solutions. Flow and heat transfer phenomena can, under continuum conditions, adequately be described in terms of dimensionless numbers such as the Reynolds, Mach, Nusselt, and Prandtl numbers However, at small pressures and with increasing gas rarefaction, the gas partly loses its continuum characteristics and the intermolecular collisions become lesser and thus macroscopic analysis is not feasible. Hence, the molecular structure of the gas needs to be taken into account. In the free molecular regime, the frequency of intermolecular collisions becomes much less and microscopic analysis is possible by using the kinetic gas theory, with the mean free molecular path and the mean molecular velocity as the important parameters. Thus the basic formulation of the flow phenomena and heat transfer in the continuum and free molecular regimes is much more refined compared to other regimes of rarefied flows such as the transitional and slip flow regimes. These remain to be understood in terms of heat transfer, and this chapter attempts to shed light on this aspect.

Traditionally the application areas for heat transfer processes in rarefied gases are space technology and missiles at high altitudes. However, during the recent years, other areas of applications have been introduced, such as

- microchannels for cooling electronic equipment
- biotechnology, e.g., cell separation

Heat Transfer in Aerospace Applications
ISBN 978-0-12-809760-1
http://dx.doi.org/10.1016/B978-0-12-809760-1.00004-1

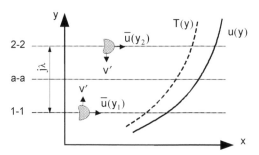

Figure 4.1 Molecular momentum exchange.

4.2 KINETIC THEORY OF GASES

Consider the flow of a gas parallel to a wall as in Fig. 4.1, with a velocity gradient and a temperature gradient normal to the wall.

Molecules pass back and forth across the arbitrary plane *a-a* parallel to the wall. During the time interval $d\tau$, molecules with the velocity v' and parallel to the plane *a-a* will pass the plane when they are in proximity $v'd\tau$ to the plane. The number of molecules per unit volume is n. The average velocity v' can be expressed as a constant times the mean molecular velocity, $i'\bar{v}$. Therefore, all molecules within the distance $i'\bar{v}d\tau$ will pass the plane *a-a*. The number of such molecules is $i'n\bar{v}d\tau$ (per unit area), or per unit time, $i'n\bar{v}$. At an arbitrary position y normal to the wall, the molecules have the velocity $\bar{u}$ (y) in the x-direction. In Fig. 4.1 when the molecules move upward from the plane 1-1, they maintain their velocity $\bar{u}$ (y_1) until they reach the plane 2-2, where they will reach the velocity $\bar{u}$ (y_2). The distance between the planes are of the magnitude of the mean free path, $j\lambda$ (λ is the mean free path). Similarly the velocities change for the downward moving molecules. The momentum exchange from this process is $m[\bar{u}(y_2) - \bar{u}(y_1)]$, where m is the mass of a single molecule. The total momentum exchange per unit area and per unit time is

$$I = i'n\bar{v}m(\bar{u}(y_2) - \bar{u}(y_1)) \tag{4.1}$$

The velocity difference $[\bar{u}(y_2) - \bar{u}(y_1)]$ can be expressed as

$$\bar{u}(y_2) - \bar{u}(y_1) = j\lambda\frac{d\bar{u}}{dy} \tag{4.2}$$

and Eq. (4.1) can be written as

$$I = i'jnm\bar{v}\lambda\frac{d\bar{u}}{dy} \tag{4.3}$$

According to Newton's second law, momentum exchange results in a force, which in this case is a shear stress. The shear stress for a Newtonian fluid reads

$$\tau = \mu \frac{d\bar{u}}{dy} \tag{4.4}$$

Eqs. (4.3) and (4.4) give

$$\mu = i'jnm\bar{v}\lambda \tag{4.5}$$

The product $nm = \rho$ is the gas density. Chapman and Enskog (see Ref. [1]) have shown that for spherical molecules, $i'j = 0.499$. According to the kinetic theory of gases, $nm\lambda = \rho\lambda$ is independent of pressure and the mean velocity of the molecules can be expressed as [1]

$$\bar{v} = \sqrt{\frac{8RT}{\pi}} \tag{4.6}$$

where R is the individual gas constant and T is the absolute temperature.

The viscosity μ can now be written as

$$\mu = 0.499\rho\lambda\sqrt{\frac{8RT}{\pi}} \tag{4.7}$$

A similar exchange between the planes will occur for thermal energy if a temperature gradient dT/dy exists.

If c_m is the specific heat per molecule, the heat exchange per unit area across the plane a-a is

$$-q = i'jn\bar{v}c_m\lambda\frac{dT}{dy} \tag{4.8}$$

Note that nc_m can be written as ρc_v, where c_v is the specific heat at constant volume, and that the thermal conductivity, k, is defined by, $-q = k \cdot dT/dy$.

Substituting these values into Eq. (4.8) gives

$$k = i'j\bar{v}c_v\rho\lambda \tag{4.9}$$

or into Eq. (4.5) gives

$$k = \mu c_v \tag{4.10}$$

If c_v is constant, k, similar to μ, will be proportional to $\sqrt{T}$.

Eq. (4.10) gives acceptable order of magnitude values but does not agree with experimental data. Chapman [1] introduced refinements in the theory that resulted in an expression for the thermal conductivity that agrees with experimental data for monoatomic molecules but not for more complex ones.

$$k = \frac{5}{2}\mu c_v \qquad (4.11)$$

Eucken [2] developed an expression for the thermal conductivity for complex molecules and it reads

$$k = \frac{1}{4}(9\gamma - 5)\mu c_v \qquad (4.12)$$

where $\gamma = c_p/c_v$.

From Eq. (4.12), the Prandtl number ($Pr = \mu c_p/k$) can be expressed as

$$Pr = \frac{4\gamma}{9\gamma - 4} \qquad (4.13)$$

which agrees well with experimental data.

4.3 FLOW REGIMES FOR RAREFIED GASES

In the analysis of flow and heat transfer processes in rarefied gases, the flow and thermal parameters must be viewed in terms of the results of the kinetic theory.

With Eqs. (4.5) and (4.6) the Reynolds number reads

$$Re = \frac{\rho UL}{\mu} = \frac{1}{0.499} \cdot \frac{U}{\bar{v}} \cdot \frac{L}{\lambda} \qquad (4.14)$$

The Reynolds number is then a product of the velocity ratio and length ratio. The velocity ratio is the ratio between the macroscopic flow velocity U and the molecular mean velocity $\bar{v}$. The length ratio describes the relation between the characteristic length L and the mean free molecular path λ.

A similar interpretation of the Mach number is also possible

$$Ma = \frac{\text{Flow velocity}}{\text{Velocity of sound}} = \frac{U}{a} = \frac{U}{\sqrt{\gamma RT}} = \frac{U}{\sqrt{\gamma \frac{\pi}{8} \bar{v}^2}} = \frac{U}{\bar{v}} \cdot \frac{1}{\sqrt{\gamma \pi/8}}$$

$$(4.15)$$

Thus the Mach number is proportional to the ratio of the macroscopic flow velocity to the mean molecular velocity.

At ordinary pressures the intermolecular distances, even in gases, are still much smaller than the body dimensions, which usually appear in heat transfer calculations, and therefore, the notion of a continuous gas is well fulfilled. However, for gases at low pressures, and correspondingly low densities, the molecular mean free path becomes comparable with the same significant body dimension, and the effects of the rarefaction phenomenon on flow and heat transfer will become significant.

The length ratio λ/L is an important parameter in the analysis of heat transfer processes in low-density gases. This ratio is referred to as the Knudsen number (Kn) and is for flow problems related to the Mach and Reynolds numbers, see Eqs. (4.14) and (4.15).

$$\text{Kn} = \frac{\lambda}{L} = \frac{1}{0.499}\sqrt{\frac{\gamma\pi}{8}}\cdot\frac{\text{Ma}}{\text{Re}} \approx \sqrt{\frac{\gamma\pi}{2}}\cdot\frac{\text{Ma}}{\text{Re}} \tag{4.16}$$

For diatomic gases $\gamma = 1.4$, the Knudsen number is

$$\text{Kn} = 1.48\frac{\text{Ma}}{\text{Re}} \tag{4.17}$$

The flow is commonly divided into different regimes depending on the Knudsen number. For boundary layer flow the characteristic dimension of greatest importance is the boundary layer thickness, δ. For laminar boundary layer flow, δ is related to the Reynolds number and the significant body dimension as

$$\delta/L \sim 1/\sqrt{\text{Re}}$$

and the corresponding Knudsen number then becomes

$$\text{Kn} = \frac{\lambda}{\delta} \sim \frac{\text{Ma}}{\sqrt{\text{Re}}} \tag{4.18}$$

Hence conventional gas dynamics will exist when $\text{Re} \gg 1$ and $\text{Ma}/\sqrt{\text{Re}} \ll 1$. It is convenient to subdivide the field of rarefied gas dynamics into several flow regimes, depending on the magnitude of the rarefaction of the gas (Fig. 4.2). The regimes are briefly described in the following.

Continuum flow. In continuum flow, the state of ordinary gas dynamics, intermolecular collisions dominate the flow field and a usual boundary condition at the interface between a fluid and a solid is that the fluid

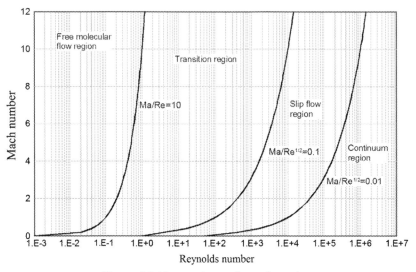

Figure 4.2 Flow regimes of gas dynamics.

adjacent to the surface assumes both the velocity and temperature of the surface. The Knudsen number for continuum flow is small, approaching zero.

$$\text{Continuum flow}: \ \frac{Ma}{\sqrt{Re}} < 0.01 \qquad (4.19)$$

Slip flow. In the near-continuum flow, where the Knudsen number is small but not negligible, one of the most interesting effects of rarefaction on the flow occurs. The gas adjacent to the surface is no longer at the velocity and temperature of the surface. The gas at the surface has a tangential velocity, it *slips* along the surface, and the temperature is different from the surface temperature, i.e., there is a *jump* in temperature between the surface and the adjacent gas. The *velocity slip* and the *temperature jump* effects are related to the molecular mean free path length, and parameters called *accommodation* and *reflection* coefficients are introduced. These describe the interaction between the statistical molecule and the surface.

$$\text{Slip flow}: \ 0.01 \leq \frac{Ma}{\sqrt{Re}} < 0.1 \qquad (4.20)$$

Knudsen flow. For extremely small densities the molecular mean free path is much larger than any significant body dimension. In such cases the molecules striking and leaving the surface do not collide with free-stream molecules until they are very far from the surface. Intermolecular collisions have a negligible effect, and the near-surface flow is characterized by the interaction between the free molecules and the surface. This flow regime is named free molecular flow or the Knudsen flow.

$$\text{Knudsen flow}: \quad \frac{\text{Ma}}{\text{Re}} > 10 \qquad (4.21)$$

Transition flow. The regime between slip flow and Knudsen flow is called transition flow and is characterized by conditions in which intermolecular collisions and collisions between gas molecules and the surface are of more or less equal importance.

$$\text{Transition flow}: \quad \frac{\text{Ma}}{\sqrt{\text{Re}}} > 0.1 \;\; \text{and} \;\; \frac{\text{Ma}}{\text{Re}} < 10 \qquad (4.22)$$

The given limits of the flow regimes are approximate and describe only the most characteristic conditions within each flow regime [3]. For example, slip flow characteristics may, in some cases, dominate above the limit given in Fig. 4.2.

4.4 METHODS OF ANALYSIS

In continuum flow, conventional methods are used, i.e., the momentum equation, mass conservation equation, and energy equation are solved. Slip flow is treated as continuum flow, but with a slip in velocity at the surface, $u_{gas} \neq u_{wall}$, and a jump in temperature, $T_{gas} \neq T_{vägg}$. For the transition regime, no thorough method of analysis has been developed; however, there exists a great deal of experimental investigations. The analytic tool for the free molecular flow is the kinetic gas theory.

4.5 INTERACTION BETWEEN GAS AND SURFACE

Establishing a real understanding requires a detailed knowledge of the physics of the interactions of molecules, with the surface distribution functions that indicate how the molecules are reflected from the surface. Such information is not yet a common knowledge.

Maxwell postulated an interaction model in which incident molecules were either reflected in a specular or a diffuse fashion. The fraction of diffusively reflected molecules is f_s, and $1 - f_s$ represents the fraction of molecules reflected in a specular manner. When both energy and momentum are of importance, three so-called accommodation coefficients are introduced:

$$\alpha_s = \frac{e_i - e_r}{e_i - e_w} \tag{4.23a}$$

$$f_s = \frac{\tau_i - \tau_r}{\tau_i} \tag{4.23b}$$

$$\sigma_s = \frac{p_i - p_r}{p_i - p_w} \tag{4.23c}$$

where e, τ, and p are the energy flux, tangential momentum, and normal momentum incident on a surface, respectively.

The indices i, r, and w represent incident on surface, reemitted from surface, and reemitted as a gas in complete Maxwellian equilibrium with the surface, respectively [3].

The coefficients f_s, σ_s, and α_s are functions of the gas composition and the gas temperature; the chemical state, physical structure, and temperature of the solid surface; and the velocity of the gas flow over the surface.

Tables 4.1 and 4.2 give the accommodation coefficient for tangential momentum and thermal energy, respectively.

4.6 HEAT TRANSFER AT HIGH VELOCITIES

Heat transfer at high velocities was analyzed in Chapter 3 but is briefly repeated here. The convective heat flux is calculated as

$$q_w = h(T_w - T_{aw}) \tag{4.24}$$

where T_{aw} is the adiabatic wall temperature and h is the heat transfer coefficient. The adiabatic wall temperature is determined by

$$T_{aw} = r \cdot \frac{U_\infty^2}{2c_p} + T_\infty \tag{4.25}$$

where r is the recovery factor.

Table 4.1 Accommodation Coefficient for Tangential Momentum f_s [4]

Gas	Surface	f_s
Air	Machined brass	1.00
Air	Old shellac	1.00
Air	Fresh shellac	0.79
Air	Oil	0.90
Air	Glass	0.89
Air	Hg	1.00

Table 4.2 Thermal Accommodation Coefficient α_s [5]

Gas	Surface	Temperature [K] $T_w \approx T_{gas}$	α_s
Air	Flat lacquer on bronze	300	0.88–0.89
Air	Polished bronze	300	0.91–0.94
Air	Machined bronze	300	0.89–0.93
Air	Etched bronze	300	0.93–0.95
Air	Polished cast iron	300	0.87–0.93
Air	Machined cast iron	300	0.87–0.88
Air	Etched cast iron	300	0.87–0.96
Air	Polished aluminum	300	0.85–0.95
Air	Machined aluminum	300	0.95–0.97
Air	Etched aluminum	300	0.89–0.97

For low-density heat transfer (high Knudsen number values) the recovery factor might be greater than that for continuum flow (Figs. 4.8 and 4.13).

4.7 SLIP FLOW REGIME

When the Knudsen number is small but not negligible, the flow field and temperature equations for continuum flow can be used in the analysis. However, at the surface the boundary conditions change because of the discontinuities appearing for velocity (slip) and temperature (jump) for the gas adjacent to the wall (Fig. 4.3).

The slip velocity and temperature jump are determined by

$$u_0 = \frac{2 - f_s}{f_s} \lambda \left(\frac{du}{dy} \right)_{y=0} \tag{4.26}$$

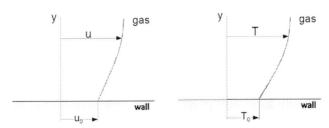

Figure 4.3 Discontinuity at the wall–gas interface in the slip flow regime.

$$T_0 - T_w = \frac{2 - \alpha_s}{\alpha_s} \; \frac{2\gamma}{\gamma + 1} \; \frac{\lambda}{Pr} \left(\frac{dT}{dy}\right)_{y=0} \qquad (4.27)$$

where $\gamma = c_p/c_v$, α_s is the thermal accommodation coefficient, f_s is the tangential momentum accommodation coefficient (Tables 4.1 and 4.2), and λ is the molecular mean free path.

Ref. [4] gives the following relation of λ, which is valid when $100 < T < 1900$ K

$$\lambda = \frac{3.1 \cdot 10^{-5} T^2}{(T + 110.4) \cdot p} \qquad (4.28)$$

where T is the absolute temperature, p is the pressure (Pa), and λ is in meters.

Eqs. (4.26) and (4.27) are valid when the temperature gradients in the tangential direction are negligible.

4.7.1 Heat Conduction in Rarefied Gases

Heat conduction across a gas layer is the result of a great number of simultaneous collisions between the gas molecules. The molecules coming from a warmer region of the gas have a higher energy content than the molecules from a colder region. The collisions between the molecules result in a more even distribution of the energy between the molecules, and in the gas a temperature gradient is established, the magnitude of which depends on the temperature of the walls surrounding the gas.

4.7.1.1 Parallel Plates

Consider Fig. 4.4 in which a gas at continuum state is situated between two parallel plates, which are separated by a distance d and with temperatures T_{w1} and T_{w2}. The heat flux is given by

$$q_c = k \frac{T_{w1} - T_{w2}}{d} \qquad (4.29)$$

As the gas pressure is reduced the molecular mean free path is increased and the importance of the molecular motion starts to become evident closest to the walls. As the Knudsen number is in the range of $0.001 \leq Kn \leq 0.1$ the gas temperature closest to the walls will differ from the wall temperature. This phenomenon is called temperature jump. The temperature distribution between the plates becomes as conjectured in Fig. 4.5. In the analysis of the heat transport, different boundary conditions

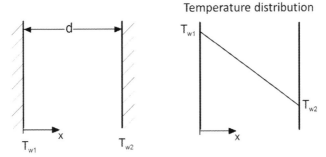

Figure 4.4 Heat conduction across a gas layer between parallel plates, in the continuum state.

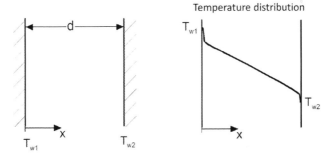

Figure 4.5 Conjectured temperature distribution, with jumps at the walls.

are needed because of the discontinuity in the temperature gradients. However, the gas away from the walls keeps its continuum character.

The effect of the temperature jump will be illustrated later in a specific example.

4.7.2 Example: Cylinder in Crossflow

The temperature equation for a flowing fluid with constant properties in cylindrical coordinates reads [4]

$$u\frac{\partial T}{\partial r} + \frac{v}{r}\frac{\partial T}{\partial \theta} + w\frac{\partial T}{\partial z} = \alpha\left(\frac{\partial^2 T}{\partial r^2} + \frac{1}{r}\frac{\partial T}{\partial r} + \frac{1}{r^2}\frac{\partial^2 T}{\partial \theta^2} + \frac{\partial^2 T}{\partial z^2}\right) \quad (4.30)$$

Consider an infinite cylinder normal to a gas flow, as in Fig. 4.6. If the flow is two dimensional, and if the radial velocity $u = 0$ and the tangential velocity $v = V$ is constant, Eq. (4.30) reduces to

$$\frac{V}{r}\frac{\partial \vartheta}{\partial \theta} = \alpha\left(\frac{\partial^2 \vartheta}{\partial r^2} + \frac{1}{r}\frac{\partial \vartheta}{\partial r}\right) \quad (4.31)$$

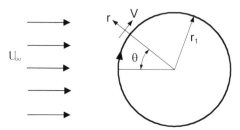

Figure 4.6 Cylinder in crossflow.

where $\vartheta = (T-T_w)/(T_{aw}-T_w)$; T_w = the constant wall temperature at the cylinder surface; T_{aw} = the adiabatic wall temperature, $T_{aw} = T_\infty + rU_\infty^2/2c_p$.

The boundary conditions are

Continuum flow:

$$r = r_1 : \vartheta = 0 \quad \text{for } 0 < \theta < \pi$$
$$r > r_1 : \vartheta = 1 \quad \text{for } \theta = 0$$

Slip flow:

$$r = r_1 : \frac{\partial \vartheta}{\partial r} = L\vartheta \quad \text{for } 0 < \theta < \pi$$
$$r > r_1 : \vartheta = 1 \quad \text{for } \theta = 0$$

where $\frac{1}{L} = 1.996 \frac{\gamma}{\gamma+1} \frac{2-\alpha_s}{\alpha_s} \frac{\lambda}{Pr}$ according to Eq. (4.27).

The solutions of Eq. (4.31) are presented by Refs. [3] and [5]. The Nusselt number for continuum flow is

$$\overline{Nu} = \frac{\overline{h}D}{k}$$

$$= 1.128 \left(\frac{K\pi}{r_1}\right)^{-0.5} + 0.500 - 0.0941 \left(\frac{K\pi}{r_1}\right)^{0.5} + 0.0625 \left(\frac{K\pi}{r_1}\right)$$

$$- 0.0588 \left(\frac{K\pi}{r_1}\right)^{1.5} + \dots$$

$$(4.32)$$

where $K = \alpha/4\,V$.

With $V = U_\infty/3$, which is an adequate and acceptable mean value for the velocity close to the cylinder surface, one obtains

$K\pi/r_1 = 3\pi/(2RePr)$. When $Re > 300$ and $Pr = 0.70$, Eq. (4.32) can be approximated as

$$\frac{\bar{h}D}{k} = 0.500 + 0.435Re^{0.5} \qquad (4.33)$$

Eq. (4.33) shows good agreement with experimental data.

The Nusselt number for slip flow is given by

$$\overline{Nu} = \frac{\bar{h}D}{k} = DL\left[1 - \frac{4}{3\sqrt{\pi}}DL\left(\frac{K\pi}{r_1}\right)^{0.5} + \frac{DL}{4}(1 + 2DL)\left(\frac{K\pi}{r_1}\right) + ...\right] \qquad (4.34)$$

With $\alpha_s = 0.8$ and $Pr = 0.7$, one has $DL \approx 0.267Re/Ma$ and, as mentioned earlier, $K\pi/r_1 = 3\pi/(2RePr)$.

Fig. 4.7 depicts results from Eqs. (4.33) and (4.34) and comparison with a bulk of experimental data from various sources. The correlation between the theoretic analysis and experimental data is quite good when $Ma < 2.0$. Fig. 4.7 has been simplified and reconstructed from a corresponding figure originally presented in Ref. [3]. Fig. 4.7 gives a clear visualization of the rarefaction effect.

4.7.3 Sphere

Consider the definition of the heat transfer coefficient for continuum flow

$$-k\left(\frac{\partial T}{\partial y}\right)_{y=0} = h^0\left(T_w^0 - T_{aw}^0\right)$$

where the superscript 0 denotes continuum condition. By combining this equation and Eqs. (4.16), (4.24) and (4.27) the Nusselt number can be written as [3]

$$\frac{Nu^0}{Nu} = 1 + \sqrt{\frac{\gamma\pi\theta^2}{2}}\frac{Ma}{RePr}Nu^0 \qquad (4.35)$$

where $\theta = 1.996\frac{2-\alpha_s}{\alpha_s}\frac{\gamma}{\gamma+1}$ and Nu^0 is the Nusselt number for continuum flow.

With air as the medium, $\gamma = 1.4$, and with $\alpha_s = 0.8$, one obtains $\frac{Nu^0}{Nu} = 1 + 2.59\frac{Ma}{RePr}Nu^0$.

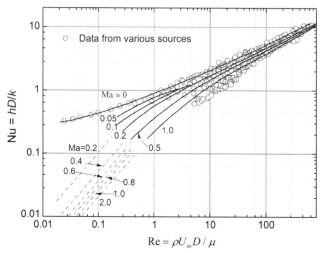

Figure 4.7 Heat transfer from a circular cylinder in crossflow, in the slip flow regime.

This equation has been correlated with experimental results (*dashed lines* in Fig. 4.9) as

$$\overline{Nu} = \frac{\overline{Nu}^0}{1 + 3.42[Ma/(RePr)]\overline{Nu}^0} \qquad (4.36)$$

Fig. 4.8 shows experimental heat transfer data for a sphere downstream a normal shock (Ma > 1) and Fig. 4.9 shows corresponding data for a subsonic flow. Fig. 4.8 has been simplified and reconstructed from a corresponding figure presented in Refs. [3] and [6]. Both Figs. 4.8 and 4.9 clearly show the effect of rarefaction.

An interesting feature of rarefied gas heat transfer is shown in Fig. 4.10. The recovery factor increases with the Knudsen number, i.e., decreasing $\sqrt{Re}/Ma$. At low Knudsen number values (high values of $\sqrt{Re}/Ma$), i.e., close to the continuum flow, the recovery factor becomes $r \approx Pr^{1/3}$.

4.7.4 Flat Plate: Tangential Flow

At laminar slip flow along a flat plate, Ref. [3] presents the following solutions for the drag coefficient and the Stanton number

$$C_D Ma = \frac{2.67}{X_1^2}\left[e^{X_1^2}erfcX_1 - 1 + \frac{2}{\sqrt{\pi}}X_1\right] \qquad (4.37)$$

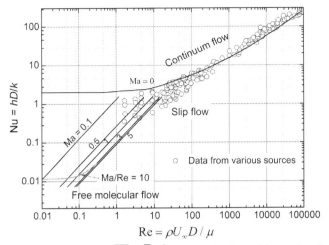

Figure 4.8 The Nusselt number ($\overline{Nu} = \overline{h}D/k$) versus the Reynolds number for a sphere downstream a normal shock [6].

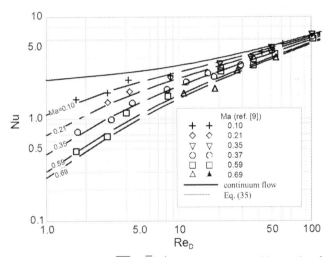

Figure 4.9 The Nusselt number ($\overline{Nu} = \overline{h}D/k$) versus the Reynolds number for a sphere at subsonic flow.

$$StMa = \frac{0.38}{X_2^2}\left[e^{X_2^2}\text{erfc}X_2 - 1 + \frac{2}{\sqrt{\pi}}X_2\right] \tag{4.38}$$

where $X_1 = \frac{2}{3}\sqrt{Re}/Ma$, $X_2 = \sqrt{Re(Pr/6.9)Ma^2}$, $St = h/\rho Uc_p$, and erfc(ξ) is the complementary error function, erfc(ξ) = $1 - \text{erf}(\xi)$.

Eq. (4.37) shows good agreement with experimental data.

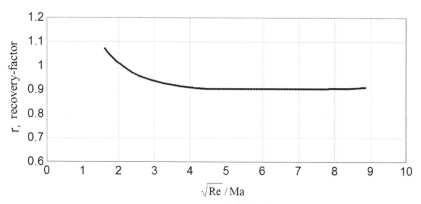

Figure 4.10 Recovery factor, $r = (T_{aw} - T_\infty)/(U_\infty^2/2c_p)$, versus the reciprocal of the Knudsen number [3].

At hypersonic flows (Ma > 5) the flow pattern becomes more complex, with interaction between shocks and the viscous boundary layer. The hypersonic flow regime can be subdivided into four characteristic regimes: (1) weak viscous interaction, (2) strong viscous interaction, (3) strong viscous interaction with surface slip, and (4) free molecular flow.

In the literature, figures showing distributions of the skin friction coefficient and the Stanton number have been presented [3,10].

4.8 TRANSITION REGIME

In the transition regime, it is recommended to use a correlation presented by Sherman [7]. The correlation is based on a dimensionless quantity F(Ma,Re), which has two limits, namely, the free molecular flow limit Re → 0 and the continuum flow limit Ma → 0.

The free molecular flow limit is defined as

$$F_{fm}(Ma) = \lim_{Re \to 0} F(Ma, Re)$$

and the continuum flow limit as

$$F_c(Re) = \lim_{Ma \to 0} F(Ma, Re)$$

Measured values of F(Ma,Re) are then plotted as F/F_{fm} versus F_c/F_{fm}, and the correlation takes the form

$$\frac{F}{F_{fm}} = \frac{1}{1 + F_{fm}/F_c} \tag{4.39}$$

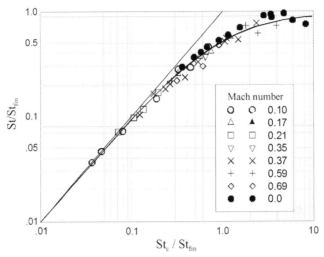

Figure 4.11 The Stanton number for a sphere at subsonic flow [3,9].

Eq. (4.39) gives a good correlation of experimental data for both the Stanton number (Fig. 4.11) and the skin friction drag coefficient.

4.9 FREE MOLECULAR FLOW REGIME: THE KNUDSEN FLOW

Free molecular flow, the Knudsen flow, occurs when the Knudsen number value is large (Eq. 4.21). Intermolecular collisions have a negligible influence, and the boundary layer flow is characterized by the interaction between free molecules and the surface. The normal and tangential momenta and the energy transfer from the surface to a rarefied gas in the Maxwellian equilibrium can be calculated entirely from the fundamental notions of the kinetic theory of gases, in which it is assumed that the so-called Maxwellian distribution function applies [8]. This assumption gives, in general, satisfactory results for engineering calculations. The details are rather complex and omitted here, see [3]. Solutions for the heat transfer, the Stanton number, and the recovery factor for a number of common geometries have been presented by Oppenheim [13] and are shown in a diagram form in Figs. 4.12 and 4.13, respectively.

4.10 EXAMPLE: LOW-DENSITY HEAT TRANSFER

Consider a thermocouple constructed by welding two thin wires together to form a spherical bead (Fig. 4.14). The bead diameter is 1 mm. The bead

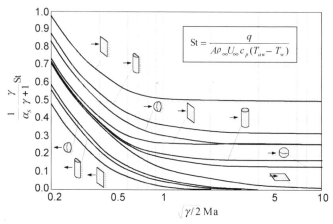

Figure 4.12 The Stanton number for different geometries at the Knudsen flow. *(Based on Eckert ERG, Drake Jr RM. Analysis of heat and mass transfer. Tokyo: McGraw-Hill; 1972.)*

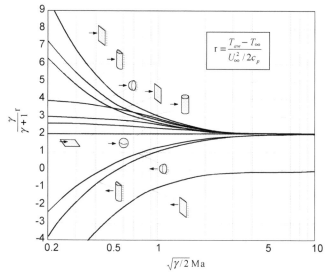

Figure 4.13 Recovery factor for different geometries at the Knudsen -flow. *(Based on Eckert ERG, Drake Jr RM. Analysis of heat and mass transfer. Tokyo: McGraw-Hill; 1972.)*

is exposed to an airstream with a velocity corresponding to $Ma = 6$ and at a pressure 10^{-6} atm and temperature $T_\infty = -70°C$. The task is to estimate the temperature of the thermocouple bead if the surface emissivity can be estimated to 0.7. It can be assumed that the accommodation coefficient of the bead is equal to that of cast iron.

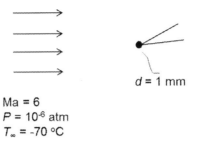

Ma = 6
P = 10^{-6} atm
T_∞ = -70 °C

Figure 4.14 Thermocouple bead of two wires.

At first, it is appropriate to determine the flow regime. The thermophysical properties are evaluated at the freestream conditions and one has $T_\infty = -70°C = 203$ K, which implies $\mu = 1.329 \cdot 10^{-5}$ kg/(m s), $k = 0.0181$ W/(m K), and $c_p = 1006$ J/(kg K).

The gas density is found from the perfect gas law as

$$\rho = \frac{p}{RT} = \frac{1 \cdot 10^{-6} \cdot 1.0132 \cdot 10^5}{287 \cdot 203} = 1.739 \cdot 10^{-6} \text{kg/m}^3$$

The velocity of sound is found from

$$a = \sqrt{\gamma RT} = (1.4 \cdot 287 \cdot 203)^{0.5} = 285.6 \text{ m/s}$$

Then one finds the freestream velocity to be $U_\infty = 6 \cdot 285.6 = 1714$ m/s. The Reynolds number is now determined and it becomes

$$Re = \frac{\rho U_\infty d}{\mu} = 0.224$$

By visiting Fig. 4.2, it is found that the free molecular range is valid ($Kn = Ma/Re = 26.8$). Then Figs. 4.12 and 4.13 can be used to determine the heat transfer parameters. The molecular speed ratio is calculated as

$$S = \sqrt{\frac{\gamma}{2}} \cdot Ma = 5.02$$

From these figures, one also finds

$$\frac{\gamma + 1}{\gamma} r = 2.04, \quad \frac{1}{\alpha} \frac{\gamma + 1}{\gamma} St = 0.127$$

By using the value of the accommodation coefficient $\alpha = 0.9$, one finds

$$St = 0.196 \text{ and } r = 1.19$$

The adiabatic wall temperature is now calculated as

$$T_{aw} = r \cdot \frac{U_\infty^2}{2c_p} + T_\infty = 1942 \text{ K}$$

From the Stanton number, St, the heat transfer coefficient h can be determined as

$$h = \rho c_p U_\infty St = 0.381 \text{ W/m}^2\text{K}$$

Now the bead surface temperature can be found by a heat balance between convection and thermal radiation as

$$h \cdot (T_{aw} - T_{surface}) = \sigma \cdot \varepsilon \cdot (T_{surface}^4 - T_\infty^4)$$

With $\sigma = 5.67 \cdot 10^{-8} \text{ W/m}^2 \text{ K}^4$, one finally obtains

$$T_{surface} = 359 \text{ K} = 86°\text{C}$$

From the results, one can easily imagine that the thermal radiation is important in low-density heat transfer cases.

4.11 EXAMPLE: HEAT TRANSFER IN AN EVACUATED SPACE

Consider the space between two vertical and parallel plates. The surface of the plates is polished aluminum, with an emissivity of $\varepsilon = 0.05$. The distance between the plates is 25 mm and air at a pressure $1.0135 \cdot 10^{-6}$ N/m^2 occupies the space between the plates. The temperature of each plate is 100 °C and 30 °C. The heat flow due to conduction across the air gap between the plates is to be calculated.

To reveal the flow regime the mean free path is first calculated (average temperature is used). The mean free path is found from Eq. (4.28) and it becomes

$$\lambda = 0.078 \text{ m}$$

The Knudsen number then becomes

$$kn = \frac{\lambda}{d} = \frac{0.078}{0.025} = 3.12$$

According to the previous regime description, it seems that the transition regime is valid, but here it is assumed that slip flow prevails. The following properties are valid:

$k = 0.0291$ W/mK, $\gamma = 1.4$, Pr $= 0.7$, $\alpha = 0.9$ (from Table 4.2).

According to Eq. (4.27) the temperature jump ΔT (at both surfaces) is found to be

$$\Delta T = \frac{2 - \alpha}{\alpha} \cdot \frac{2\gamma}{\gamma + 1} \cdot \frac{\lambda}{Pr} \cdot \frac{T_{w1} - T_{w2}}{d}$$

which will give

$$\Delta T = 32.4°C$$

Then the conductive heat transfer between the plates is

$$q = k \frac{T_{w1} - T_{w2} - 2\Delta T}{d} = 6.1 \text{ W/m}^2$$

At normal atmospheric pressure the corresponding heat flux will be

$$q = k \frac{T_{w1} - T_{w2}}{d} = 81.5 \text{ W/m}^2$$

It is thus evident that the low-density effect is large.

By considering the radiative heat transfer between the two parallel surfaces, one has [4]

$$q_{rad} = \frac{\sigma\left(T_{w1}^4 - T_{w2}^4\right)}{\dfrac{2}{\varepsilon} - 1} = 22.5 \text{ W/m}^2$$

Thus it is obvious that the thermal radiation is much higher than the conduction heat transfer at low-density conditions, even if the surfaces are highly polished ($\varepsilon = 0.05$).

4.12 MICROCHANNEL APPLICATIONS

Several studies have been conducted on rarefied gas dynamics in microchannels. This is because of the developing interest on microscale heat transfer for applications in electronic equipment, micro heat exchangers, sensors and flow controls, and microelectromechanical systems (MEMSs). Some papers and review have been presented [11–17]. As mentioned earlier, the continuum approach breaks down for large mean free paths as in rarefied gas dynamics. The Knudsen number was introduced and it provides a direct mean of validating the continuum approach as it compares the mean free path λ to the characteristic length d. The flow regimes defined in relation to the Knudsen number were introduced previously. Gaseous flow and heat transfer in micro-and minichannels have been extensively studied because of the importance in development and design of microdevices in response to the need for thermal control in the fabrication and operation of micro- and nanoscale devices such as high-speed applications, high-density microscale electronic devices, microsensors, and micromachines. In addition to meeting microscale requirements, micro- and nanoscale cooling devices must be capable of extremely high performance. The rarefied gas flow dynamics and heat transfer are also important in vacuum and space technology. However, several contradictory results have been published in terms of pressure drop and heat transfer coefficients. Also experiments have been performed to validate theoretic or numerical results.

For small values of the Knudsen number, the continuum theory is applicable and the macroscopic equations for conservation of mass, momentum, and energy can be used to describe the flow and heat transfer. For very large values of the Knudsen number (free molecular regime), the number of molecules colliding with the walls becomes larger than that of the intermolecular collisions. The molecules move independently from each other and transport energy and momentum directly from one wall to another. In the kinetic gas theory the flow and heat transfer processes are described by the Boltzmann equation [18–20]. The Boltzmann equation is an integrodifferential equation.

In the kinetic gas theory, the thermodynamic state of a gas is described with the molecular velocity distribution function. This function describes the statistical distribution of the molecular velocities and has to be found as a solution of the Boltzmann equation. The thermodynamic variables are then determined as moments of the molecular velocity distribution function. For a gas in thermodynamic equilibrium, the Maxwellian distribution is valid.

There are several approaches to solve the Boltzmann equation. The suggested approaches are expansions of the distribution function about the absolute thermodynamic equilibrium or about the local thermodynamic equilibrium and replacing the nonlinear collision term of the Boltzmann equation by a simple mathematic model or solution of the momentum equations. Numerical simulations based on the Monte Carlo method or molecular gas dynamics have been performed. Solutions of the Boltzmann equation for problems with simple geometries such as parallel plates, concentric cylinders, and spheres have been presented. For polyatomic gases in engineering applications the Boltzmann equation is generally too complicated and instead available solutions are based on empirical assumptions. Solutions for monoatomic gases are modified by introducing macroscopic thermodynamic properties of polyatomic gases.

4.12.1 The Direct Simulation Monte Carlo Method

The direct simulation Monte Carlo (DSMC) method of gaseous flow and heat transfer was originally developed by Bird to solve problems related to flow around space vehicles at high altitudes. The DSMC method can be described as a probability/statistical simulation method that utilizes a sequence of random numbers to perform the simulation. Unlike computational fluid dynamics (CFD), the DSMC method simulates the physical process directly without performing a mathematic discretization of the partial differential equations that describe the mathematic system. However, the DSMC method requires that the system is described by a probability density function. As this requirement is fulfilled, the simulation is performed by a random sampling of this function. The random sampling also requires generation of a random number, Rf, which is uniformly distributed between 0 and 1. The output of the simulation is then accumulated at each sampling and averaged over the number of samples to arrive at the solution of the physical problem. The DSMC method has been very useful in gas dynamics, particularly for dilute gases. Bird [21] developed a DSMC code to perform molecular simulations. The gas is modeled by a large number of simulated molecules, which range in several millions. Each simulated molecule represents a large number of real molecules. The information about the simulated molecules (position in space, velocity components, internal state) is saved and updated with time after representative collisions and boundary interactions in the simulated physical space. The simulation follows the trajectory of a very large number of simulated molecules. There is neither a convergence criterion for the DSMC method nor a requirement for an initial approximation of the flow and temperature fields. However, there are numerical criteria that has to be satisfied by the DSMC method: the ratio of the time step to the local collision time and the ratio of the mean separation between collision partners to the local mean free path, which must be very small compared to unity over the entire flow field. In addition, the cell size should be chosen such that its linear length is small in comparison with the local mean free path of the molecules.

The DSMC algorithm consists of four primary processes, namely, molecular movements, indexing and cross-referencing of molecules, simulating the collisions, and sampling the flow field. Before simulation the domain is divided into a number of cells that are utilized to sample the macroscopic properties. The cells are further discretized into subcells that are used to select possible collision pairs. In addition, the number of real molecules in the gas is approximated by a lesser number of molecules called simulated molecules because of memory constraints as well computational processing time considerations. During the simulation these four processes are uncoupled using a time step that is much less than the mean collision time.

The molecular movement process is modeled deterministically and the simulation follows on a fixed spatial grid that defines the spatial cells. Modeling the molecule—surface interactions requires application of conservation laws to individual molecules. The indexing and tracking processes are the prerequisite for modeling collisions and sampling the flow field. Molecular collision is a probabilistic process that sets the DSMC method apart from deterministic simulation methods, such as molecular dynamics. Several collision modeling techniques have been applied to the DSMC method and the current preferred method is the no-time-counter technique. The final process is sampling the macroscopic flow properties. The spatial coordinates and velocity components of molecules in a particular cell are used to calculate macroscopic quantities at the geometric center of the cell. The pressure and the shear stress at the walls are calculated from the momentum exchange between the simulated molecules and the boundaries according to a surface model (diffuse, specular, or a combination of both). More details of the DSMC methodology are documented by Bird [19] and Sayegh et al. [16].

Xian et al. [15] performed an analysis and simulation of rarefied nitrogen gas flow and heat transfer for the Knudsen number values ranging from 0.05 to 1.0 by using the DSMC method. It was found that the Knudsen number and the microchannel aspect ratio greatly influenced the heat transfer performance. The microchannel inlet and outlet had higher heat fluxes than the middle part of the channel. It was found that the inlet freestream flow velocity had a minor effect on the wall heat flux but it changed the distribution of the local heat flux. The gas temperature slip on the channel surface increased by increasing the Knudsen number.

As a flow field is far from local thermodynamic equilibrium or when surface effects are dominant, local regions of noncontinuum or nonequilibrium flow may appear. In such cases, molecular simulation tools can provide accurate modeling opportunities but usually the computational efforts become huge as engineering spatial and temporal scales have to be resolved. Multiscale computational methodologies have therefore been developed to separate the scales. Multiscale methods combining continuum and molecular descriptions of the flow are called hybrid methods. Recently, Docherty et al. [22] proposed a heterogeneous multiscale method where the DSMC method was used for the molecular description. The method was demonstrated for simple energy transfer problems.

REFERENCES

[1] Chapman S, Cowling TG. The mathematical theory of non-uniform gases. 2nd ed. London: Cambridge University Press; 1958.

[2] Eucken A. On the thermal conductivity, specific heat and the viscosity of gases (in German). Phys Z 1913;14:324−32.

[3] Eckert ERG, Drake Jr RM. Analysis of heat and mass transfer. Tokyo: McGraw-Hill; 1972.

[4] Sundén B. Introduction to heat transfer. UK: WIT Press; 2012.

[5] Carslaw HS, Jaeger JC. Conduction of heat in solids. 2nd ed. New York: Oxford University Press; 1959.

[6] Drake RM, Backer GH. Heat transfer from spheres to a rarefied gas. Trans ASME 1952;74(7).

[7] Sherman FS. A survey of experimental results and methods for transition regime of rarefied gas dynamics. Adv Appl Mech 1963;2(Suppl. 2):228−60.

[8] Rohsenow WM, Hartnett JP, Ganic EN. Handbook of heat transfer fundamentals. McGraw-Hill; 1985.

[9] Kavanau LL. Heat transfer from spheres to a rarefied gas in subsonic flow. Amazon, Paperback Book; 1953.

[10] Oppenheim AK. Generalized theory of convective heat transfer in free-molecule-flow. J Aeron Sci 1951;18(5):353−4.

[11] Islam T. Rarefaction effects on the flow characteristics in microchannels on asymmetric wall thermal condition. Int J Appl Math Res 2015;4(1):119−28.

[12] Shokouhmand H, Bigham S, Isfahani RN. Effects of Knudsen number and geometry on gaseous flow and heat transfer in a constricted channel. Heat Mass Trans 2011;47:119−30.

[13] Kandlikar SG, Colin S, Peles Y, Garimella S, Pease RF, Btarnadner JJ, et al. Heat transfer in microchannels-2012 status and research needs. ASME J Heat Trans 2013;135. paper no 091001.

[14] Bayazitoglu Y, Tullius JF. Single-phase gaseous flows in microchannels. In: Encyclopedia of microfluidics and nanofluids. New York: Springer Science Business Media; 2014.

[15] Xian W, Wang QW, Tao WQ, Zheng P. Simulation of rarefied gas flow and heat transfer in microchannels. Sci China Ser E Technol Sci 2002;45(3):321−7.

[16] Sayegh R, Faghri M, Asako Y, Sunden B. Direct simulation Monte Carlo of gaseous flow and heat transfer in a microchannel [Chapter 7]. In: Faghri M, Sunden B, editors. Heat and fluid flow in macroscale and nanoscale structures. UK: WIT Press; 2004. p. 273−302.

[17] Lewandowski T, Ochrymiuk T, Czerwinska J. Modeling of heat transfer in micro-channel gas flow. ASME J Heat Trans 2011;133:022401.

[18] Bird GA. Molecular gas dynamics, Oxford Engineering Science. Oxford: Oxford University Press; 1976.

[19] Bird GA. Molecular gas dynamics and the direct simulation of gas flows. Oxford: Clarendon Press; 1994.

[20] Frohn A, Roth N, Anders K. Heat transfer and momentum flux in rarefied gases. VDI Heat Atlas 2010;M10:1375—90.

[21] Bird GA. The DS2G program user's guide. version 3.1. Killara, Australia: G.A.B. Consulting Pty Ltd; 1998.

[22] Docherty SY, Borg MK, Lockerby DA, Reese JM. Multiscale simulation of heat transfer in a rarefied gas. Int J Heat Fluid Flow 2014;50:114—25.

CHAPTER 5

Cryogenics

5.1 INTRODUCTION

This chapter presents applied heat transfer principles in the range of extremely low temperatures. The specific features of heat transfer at cryogenic temperatures, such as variable properties, near-critical convection, and the Kapitza resistance, are described. The chapter will include some examples to illustrate specific phenomena.

The cryogenic temperature range is commonly defined as from $-150°C$ to absolute zero (i.e., $-273°C$ or 0K). At 0K the molecular motion comes as close as theoretically possible to ceasing completely. Cryogenic temperatures are considerably lower than those encountered in ordinary physical processes. At these extreme conditions the thermal conductivity, ductility, strength of materials, and electric resistance are altered, which is of great importance. As heat is considered to be created by the random motion of molecules, this implies that substances at cryogenic temperatures are very close to a static and highly ordered state. The development of cryogenics has been connected to the development of refrigeration systems. As the ability of many supercooled metals to lose resistance to electricity was discovered, the phenomenon of superconductivity was introduced. Temperatures below 3K are primarily used in laboratories, particularly in the research of the properties of helium. Helium is liquefied at 4.2K and this liquid is called helium I. At 2.17K, it is abruptly transferred to helium II. This liquid has such a low viscosity that it can crawl up the side of a glass and flow through microscopic holes too small to permit the passage of ordinary liquids, including helium I. This property is called superfluidicity.

A very important commercial application of cryogenic liquefaction methods is storage and transportation of liquefied natural gas (LNG). Natural gas is liquefied at 110K and becomes very compact and accordingly suitable for efficient transport in specially insulated tankers or containers. For preservation of food, very low temperatures are also used. The food products are placed in well-sealed tanks and liquid nitrogen (LN_2, liquefied at 77K) is sprayed over them. The nitrogen vaporizes immediately and absorbs the heat content of the products.

Heat Transfer in Aerospace Applications
ISBN 978-0-12-809760-1
http://dx.doi.org/10.1016/B978-0-12-809760-1.00005-3

In the so-called cryosurgery, scalpels or probes are cooled by LN_2 and then used to freeze unhealthy tissue. Then the dead cells are removed by normal processes. An advantage of cryosurgery is that freezing the tissue results in less bleeding when compared to cutting methods.

For space vehicles, cryogens such as liquid hydrogen (LH_2) and liquid oxygen (LOX) are used as propellants. The major disadvantage of using hydrogen as fuel in aerospace vehicles is its need for a large storage volume. At standard pressure and temperature, hydrogen has a density of about $0.09 \ kg/m^3$, whereas gasoline and kerosene have about $800 \ kg/m^3$. This is the main reason why hydrogen is stored under cryogenic conditions (20.46K) in liquid phase. At a given amount of energy the volume of hydrogen would be four times larger than that of kerosene. LH_2 is also proposed to have a dominant role in clean and fossil-free energy systems in transportation.

The superconductivity of materials being cooled to an extremely low temperature has a significant application in the construction of super-conducting electromagnets for particle accelerators. Such large research facilities require very powerful electromagnetic fields.

In big cities, it is difficult to transmit electric power by overhead cables and hence, underground cables are used extensively nowadays. However, the underground cables become heated and the resistance of the wires increases, leading to waste of power. Superconductors are frequently used to increase the power throughput, which requires cryogenic liquids such as nitrogen and helium to cool special alloy-containing cables to increase the power transmission.

Cryocoolers (heat exchangers operating at very low temperatures) are required on many satellites to cool infrared and microwave detectors, and thus, sharper images can be received.

5.2 KAPITZA RESISTANCE

The Kapitza resistance is a thermal resistance to heat transfer across the interface between liquid helium and a solid. In liquid helium and solids (e.g., copper), heat is carried by phonons, which are thermal equilibrium sound waves with frequencies in the gigahertz and terahertz region. The acoustic impedance of helium and solids can differ up to 1000 times, which implies that the phonons mostly reflect at the boundary, like an echo from a cliff face. This property, together with the fact that the number of phonons decreases very rapidly at low temperatures, means that at about 1K, there are few phonons to carry heat and even fewer to get across the interface. The prediction is that the Kapitza resistance at the interface is comparable to the thermal resistance of a10-m-long copper material with the same cross section.

Significant interest has been shown in the thermal properties of amorphous polymers at low temperatures. Such polymers, such as Kapton, exhibit good mechanical, chemical, and electrical properties. Therefore, they are used in many cryogenic applications such as in thermal and electrical insulation for superconducting magnet winding, as key components for cryogenic target or space applications, and in low-temperature heat exchangers. For all these cryogenics applications an accurate design is required, and thus the knowledge of the thermal properties of such materials, such as thermal conductivity and thermal resistance, dominated by the Kapitza resistance, at the solid—He II interface is essential. For pressurized helium II the reciprocal Kapitza resistance, i.e., conductance is about 0.6 W/cm^2K. However, it has been found that the impedance mismatch between a solid and a cryogen is dominant at very low temperatures and small above approximately 4K, and accordingly, it is important only for helium systems.

5.2.1 Kapitza Number

The Kapitza number (Ka) is a dimensionless number expressing the ratio of the surface tension force to inertial force. Physically it indicates the hydrodynamic wave regime in falling liquid films. In evaporators, heat exchangers, microreactors, absorbers, electronic and microprocessor cooling, and air-conditioning, liquid films are important. The Kapitza number is a material property and is defined as

$$Ka = \frac{\sigma}{\rho(g \sin \beta)^{1/3} v^{4/3}}$$

where σ is the surface tension (N/m); g, the gravitational acceleration (m/s^2); ρ, the density (kg/m^3); β, the inclination angle (radian); and v, the kinematic viscosity (m/s^2).

5.3 CRYOGENIC TANKS

Cryogens, in particular, LH_2 and LOX, are important as both power supply and life support fluids in space explorations because of their high-efficient thrust and nonpolluting waste [1,2]. During rocket launch, cryogenic propellant tanks are exposed to severe aerodynamic heating and different space radiations. As the liquid propellants have a low boiling point, they are highly sensitive to heat leaks from the external thermal environment [3]. Complex heat and mass transfer exchanges are involved in the pressurization and thermal stratification process in cryogenic tanks. The heat will be carried to the liquid—vapor interface by conduction and natural convection causing vaporization, which in a closed tank results in pressurization. Thus it is essential to understand and be able to control the pressurization and thermal stratification processes in cryogenic tanks, and this will be important for successful drainage and safe storage of cryogenic propellant for a long time, as well as for successful operation of space emissions.

Several experimental and numerical investigations have been performed on the ground under normal gravity and some under microgravity. For instance, experimental investigations of LH_2 stratification have been conducted. Effects of side-wall heating and bottom-wall heating have been studied. Various theoretic studies have been conducted but the more recent ones are based on computational fluid dynamics (CFD). Reviews can be found in Refs. [4—7]. In aerospace applications, cryogenic tanks will experience orbital transfer, slow rotation, and altitude adjustment during the coast period, and accordingly, fluid thermal stratification during such conditions has been considered in Ref. [8].

5.4 ANALYSIS OF PRESSURIZATION AND THERMAL STRATIFICATION IN AN LH_2 TANK

This section presents an analysis of self-pressurization and thermal stratification in a closed LH_2 tank by employing the so-called volume of fluid (VOF) model and phase change effect. The phenomena are investigated in a partially filled LH_2 tank for different fill levels and in-leak heat fluxes. A full tank and a partially filled tank are investigated.

5.4.1 Mathematical Model

An axisymmetric cylindrical tank partially filled with LH_2 is shown in Fig. 5.1. Only hydrogen vapor is considered to be present in the vapor space.

The VOF method [9], which is a kind of Eulerian method, has been widely used in predicting various two-phase fluid flows and is adopted in this analysis. The VOF formulation relies on the fact that two or more fluids

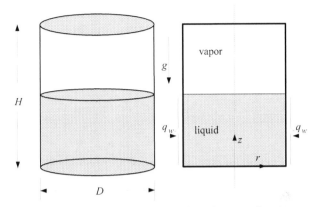

Figure 5.1 Cylindrical cryogenic tank under consideration.

do not interpenetrate each other. For each phase considered in the model a variable is introduced as the volume fraction of the phase in the computational cell. In each of the control volumes, the volume fractions of all phases sum up to unity. Because the temperature changes slightly, all the fluid properties, except density, are considered constant.

The density variation versus temperature is described by the Boussinesq approximation for evaluation of the buoyancy force. In the solution procedure the governing equations for conservation of mass, momentum, and energy are given by

$$\frac{\partial \rho}{\partial \tau} + \nabla \cdot \left(\rho \vec{V} \right) = 0 \tag{5.1}$$

$$\frac{\partial \left(\rho \vec{V} \right)}{\partial \tau} + \nabla \cdot \left(\rho \vec{V} \vec{V} \right) = - \nabla p - \rho \beta \vec{g} (T - T_0)$$
$$+ \nabla \cdot \left[\mu_{\text{eff}} \left(\nabla \vec{V} + \nabla \vec{V^T} \right) \right] + \vec{F} \tag{5.2}$$

$$\frac{\partial \rho E}{\partial \tau} + \nabla \cdot \left(\vec{V} (\rho E + p) \right) = \nabla \cdot (k_{\text{eff}} \nabla T) + S_h \tag{5.3}$$

where $\vec{F}$ is the body force resulting from surface tension at the interface and S_h is the energy source related to the phase change. A formulation of the continuum surface force (CSF) model is used, and the surface tension can finally be written in terms of the pressure jump across the interface surface. The force at the surface can be expressed as a volume force F_{vol}, using the divergence theorem [10]. The volume force acts as the source term in the momentum equation and has the following form:

$$F_{vol} = \sigma_{lv} \frac{\alpha_l \rho_l \kappa_v \nabla \alpha_v + \alpha_v \rho_v \kappa_l \nabla \alpha_l}{0.5(\rho_l + \rho_v)} \tag{5.4}$$

where σ_{lv} is the interfacial surface tension between the liquid and the vapor. The curvatures of the liquid and vapor are defined as

$$\kappa_l = \frac{\nabla \alpha_l}{|\nabla \alpha_l|}, \kappa_l = \frac{\nabla \alpha_v}{|\nabla \alpha_v|} \tag{5.5}$$

The tracking of the interface between the phases is accomplished by solving the continuity equation for the volume fraction of the second phase. This equation reads

$$\frac{\partial}{\partial \tau}(\alpha_v \rho_v) + \nabla \cdot (\alpha_v \rho_v \vec{V}) = \dot{m} \tag{5.6}$$

where $\dot{m}$ is the phase change flow rate caused by evaporation or condensation at the interface, positive or negative, respectively. It is regarded as a source term in the continuity equation. The Lee phase change model [11] is applied to consider this mass transfer and is described as

$$\dot{m} = r_l \alpha_l \rho_l (T_l - T_{sat}) \quad \text{if } T_l \geq T_{sat} \tag{5.7a}$$

$$\dot{m} = r_v \alpha_v \rho_v (T_v - T_{sat}) \quad \text{if } T_v \leq T_{sat} \tag{5.7b}$$

The saturation temperature, T_{sat}, changes according to the pressure in the tank by the Clausius–Clapeyron equation. The coefficient, r, is determined by the trial and error procedure described in Ref. [12]. Thus the energy source term S_h will be

$$S_h = L_H \dot{m} \tag{5.8}$$

where L_H is the latent heat of hydrogen.

The properties appearing in the transport equations are determined by the presence of the component phases in each control volume. For instance, thermal conductivity, density, dynamic viscosity, and specific heat can be determined by the following respective expressions:

$$k = \alpha_l k_l + (1 - \alpha_l) k_v \tag{5.9}$$

$$\rho = \alpha_l \rho_l + (1 - \alpha_l) \rho_v \tag{5.10}$$

$$\mu = \alpha_l \mu_l + (1 - \alpha_l) \mu_v \tag{5.11}$$

$$c_p = \frac{1}{\rho} [\alpha_l \rho_l c_{pl} + (1 - \alpha_l)] \rho_v c_{pv} \tag{5.12}$$

5.4.2 Thermal Environment

The cryogenic tank is exposed to serious aerodynamic heating and various heat leaks, and all the environmental physical properties change with height. Aerodynamic heating is maximum in the atmosphere, but in space the cryogenic tank is subjected to various space heat leakages. Solar incident radiation, earth albedo radiation, infrared radiation, and deep-space infrared radiation are the main heat sources. The method to estimate these four kinds of space radiation can be found in Ref. [3]. In the analysis in this chapter the heat flux is given certain values and presents effective mean values to enable parametric analysis.

The boundary conditions are as follows, see Fig. 5.1. The top and bottom surfaces of the tank are assumed to be flat and perfectly insulated, respectively. Heat-in-leak takes place only at the cylindrical wet walls and different rates of the heat-in-leak q_w are specified. The heating of LH_2 at the walls induces free convection currents, with the warmer LH_2 in the near-wall region being transferred to the upper regions of the liquid column.

Quiescent saturation conditions are assumed to prevail before the heat flux q_w is imposed at the cylindrical walls. The initial conditions at $\tau = 0$ are

$$u(r, z) = u_r = u_z \tag{5.13}$$

The initial pressure is set to 1 atm (101.32 kPa) and the initial temperature corresponds to the H_2 saturation temperature at that pressure (20.268 K). The initial temperature is assumed to be the same throughout the liquid and vapor. The pressure in the liquid is taken as a function of the height and density. No slip boundary conditions are imposed on the sidewalls.

The top and bottom surfaces are assumed to be insulated and an adiabatic boundary condition is thus valid:

$$\frac{\partial T}{\partial n} = 0 \tag{5.14}$$

whereas on the sidewalls the Neumann boundary condition is formulated as

$$-k\frac{\partial T}{\partial n} = q_w \tag{5.15}$$

5.4.3 Numerical Solution Procedure

The tank has a diameter of 0.5 m and its height is 1 m. Because of the cylindrical geometry, imposed boundary conditions, and the physics of the problem, symmetry conditions can be applied. Thus the flow and temperature fields are treated as axisymmetric. The influence of the fill level on the pressurization and thermal stratification is analyzed for three fill levels of 30%, 50%, and 80%. Various values of the incident heat flux are, respectively, considered: 50, 150, and 250 W/m². For the lowest wall heat flux the influence of the fill level is presented. If gravity is present, an important parameter for determining the flow regime is the dimensionless Rayleigh number. It is defined as

$$Ra^* = \frac{g\beta\rho^2 c_p q_w L^4}{\mu k^2} \tag{5.16}$$

For a heat flux of 50 W/m² the Ra is $1.13 \cdot 10^{13}$.

Sometimes the so-called Bond number, Bo, is introduced to reveal the ratio of buoyancy forces to surface tension. It is defined as

$$Bo = g(\rho_l - \rho_v)L^2/\sigma \tag{5.17}$$

For zero gravity, $Bo = 0$ and the surface tension becomes very important, whereas for normal gravity, the Bo is quite high and the surface tension is less important.

The commercial CFD code ANSYS FLUENT was used as a solver for the conservation equations. The interfacial mass and heat transfer model was implemented via the so-called user-defined functions.

The SIMPLEC (semi-implicit method for pressure linked equations consistent) procedure was chosen as the pressure–velocity coupling algorithm. For the pressure interpolation required to solve the momentum equation, the body-force-weighted scheme was applied because it is effective in numerically solving the buoyant natural convection problems. A second-order upwind scheme was chosen for the convection terms in the conservation equations. For turbulent cases, the k–ε turbulence model was selected based on some initial tests. The enhanced wall function approach is used for handling the near-wall region. The y^+ values closest to the solid walls were within the recommended range for this turbulence model. The maximum y^+ value was 30. Turbulent flow prevailed only under normal gravity. As the Rayleigh number is high, no particular model for handling the transition from laminar to turbulent flow was adopted.

The geometric reconstruction scheme using the piecewise linear approximation is applied for the volume fraction equation to capture the interface.

Sensitivity tests were carried out to reveal the importance of the number of control volumes of the computational grid. A grid with a total of 20,000 quadrilateral elements was used for the axisymmetric simulations, with successively increasing number of control volumes toward the walls and interface. As the computations concerned transient heat transfer and fluid flow in the domain, the time step had to be chosen in such a way that the Courant number is less than 0.1.

It should be noted that with the VOF method, it is possible to identify the various regions being occupied by the vapor and liquid, but the detailed mechanisms of the phase change process cannot be detected.

In visualization and interpretation of the flow field, the stream function is a convenient property to consider. It is generally formulated as a relation between the streamlines and the statement of conservation of mass. A streamline is a line that is tangent to the velocity vector of the flowing fluid.

In this case, as an axisymmetric flow is considered, the stream function is defined as

$$\rho u_r = \frac{1}{r}\frac{\partial \psi}{\partial z}, \rho u_z = -\frac{1}{r}\frac{\partial \psi}{\partial r} \qquad (5.18)$$

where u_r and u_z are the radial and axial velocities, respectively.

Further details are available in Ref. [13].

5.4.4 Results

In this section, simulations of self-pressurization and thermal stratification in a partially filled LH_2 tank are presented. When the influence of the imposed heat flux is investigated, the comparison is carried out at a fixed fill level. Similarly, the imposed heat flux is fixed as the fill level effect is studied. The influence of gravity, surface tension, and wettability will be presented in Chapter 9.

Fig. 5.2 shows the pressure rise versus time in the LH_2 tank for various values of the sidewall heat flux. The pressure starts to rise after a certain time. During this period, there is only marginal evaporation. This can be attributed to a coupling between the buoyancy force and convective cooling. Warm liquid created by the sidewall heat flux needs some time to flow and reach the interface, and there is also increase in the enthalpy. At the onset of evaporation the pressure rises gradually and the rise rate tends to gradually approach a constant value. The mass of the LH_2 left in the tank decreases as the evaporation process progresses. A similar phenomenon was also found by Kumar et al. [14] by using a homogeneous two-phase flow

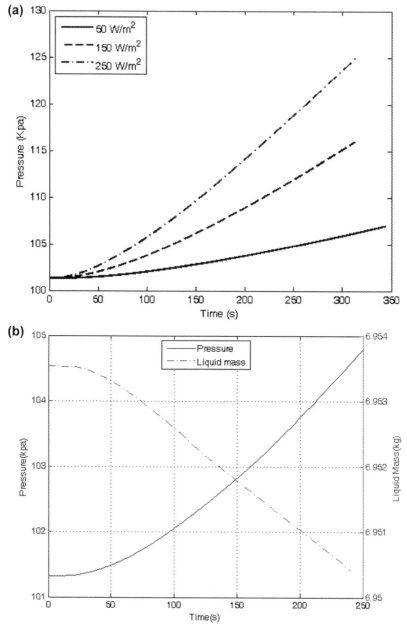

Figure 5.2 (a) Variation of vapor pressure for different heat fluxes. (b) Variation of vapor pressure and liquid mass with time. The heat flux is 50 W/m².

model. As the heat flux increases, the elapsed time becomes shorter before the evaporation is initiated. This is because the buoyancy force is essentially proportional to the difference between the temperatures of the warm liquid and the interface. The timescale for the liquid enthalpy increase and the liquid motion to the interface becomes shorter. However, as the heat flux is increased, more liquid is initially evaporated and a longer initial transient period is observed. This was also found in the experiments by Seo and Jeong [15]. After the initial evaporation phase, the liquid approaches a stable configuration, even though the average temperature is still increasing. A certain temperature stratification appears in the liquid. A part of the thermal energy entering through the tank wall is used to raise the average liquid temperature and the remaining is transferred to the vapor region. Both parts resume constant values as a stable configuration is reached. Eventually, the rate of the pressure rise becomes constant. As the heat flux increases, the pressure and pressure rise rate increase.

Fig. 5.3 presents temperature distributions in the LH_2 tank after 100 s of heating. It can be observed that temperature stratification appears in the liquid zone. The stratification is more intensive for larger heat fluxes. At the sidewall where the heat load is imposed, steep temperature gradients occur. This also happens at the interface (Fig. 5.3a and b). The temperature at the interface is higher than the given temperature range for the heat flux

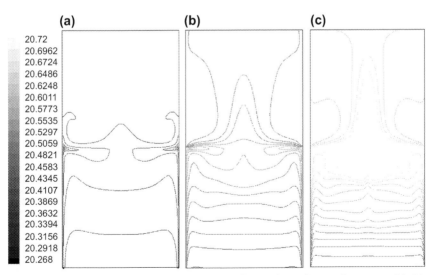

Figure 5.3 Temperature contours for three different heat fluxes after 100 s: (a) 50, (b) 150, and (c) 250 W/m^2.

$250 \ W/m^2$, so it is not shown in Fig. 5.3c. As the sidewall heat flux is increased, not only does the free surface temperature increase but also evaporation commences before a stratified layer is built up. In the vapor part, temperature stratification occurs mainly in the radial direction with a steep gradient at the centerline. This can be explained by the fluid flow, as shown in Fig. 5.4.

Fig. 5.4 shows that the liquid near the heated wall moves up because of buoyancy forces. After reaching the top surface, it turns toward the central part. Because of the symmetry at the centerline, it then turns toward the bottom. Circulating flow patterns are formed in the liquid. The downward movement near the wall indicates that the circulating fluid temperature is higher than that of the layer adjacent to the boundary layer along the wall. This is because in case of a uniform wall heat flux, the wall temperature increases with height [16], and the circulating fluid is hottest when it reaches the top of the cavity. Although the fluid loses some heat while executing its first loop, its temperature may still be higher than that of the boundary layer adjacent to the wall. A secondary loop is also observed. Below this circulating zone, there is a bulk fluid movement toward the bottom to compensate the upward flow near the heated walls. This kind of fluid flow was also observed in experiments carried out by Das et al. [17]. The vapor part seems to be heated from the bottom by the evaporation process. The hottest liquid parts meet at the center of the interface. Part of

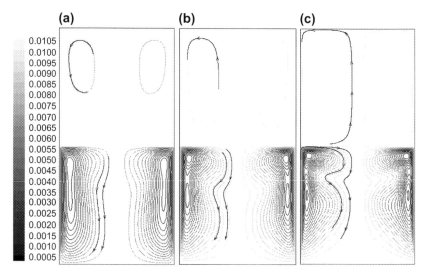

Figure 5.4 Streamlines for three different heat fluxes at 100 s: (a) 50, (b) 150, and (c) $250 \ W/m^2$.

the liquid enters the flow loops in the liquid region for cooling, whereas another part becomes vapor by evaporation, and the latent heat of the phase change is absorbed. The heated vapor in the central part reaches the top of the tank because of the buoyancy forces. After the hot vapor has reached the top, it turns toward the cool side vapor wall. For the smaller heat flux, more time is needed to develop the vapor flow. At the same time, the larger the heat flux, the bigger the developed flow region. As time elapses, the pressure rise rate becomes constant and the liquid region represents a thermally stable and stratified region (Fig. 5.5). The contour plots are combined into a single image with the isotherms to the left and the streamlines to the right in each of Fig. 5.5(a)—(c). The average temperature of the liquid becomes high and the liquid remaining for cooling is not sufficient, so the hot liquid flow intensively induces many small loops just below the interface.

In Fig. 5.6 the tank pressurization behavior is depicted for different fill levels. The pressure variations for the fill levels 30%, 50%, and 80% at a heat flux of 50 W/m^2 are shown. Again the pressure rise begins after a certain period of sidewall heating. The liquid is heated and this forces the enthalpy to exceed the saturation value corresponding to the vapor hydrogen pressure. Evaporation is then assumed to occur. The superheated period decreases as the fill level increases. A longer time is required before the pressure rise rate becomes constant for higher fill levels. The larger the liquid fraction, the larger the pressure rise caused by the expansion of

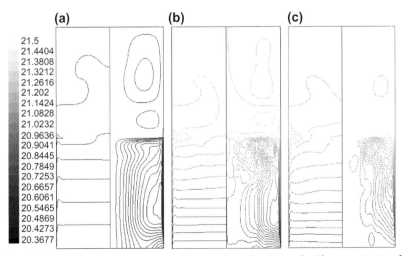

Figure 5.5 Isocontours of temperature and stream function of self-pressurization for three different heat fluxes: (a) 50, (b) 150, and (c) 250 W/m^2.

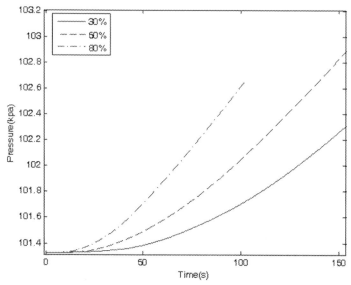

Figure 5.6 Variation of vapor pressure for different fill levels.

the initial liquid volume. The thermal stratification degree in the liquid is similar, so about the same amount of heat is transferred to the vapor. The vapor volume is smaller for high fill levels and accordingly the pressures are higher than those for low fill levels.

The temperature variation and stream function for three fill levels are illustrated in Fig. 5.7. It is observed that the stratification degree is similar, and the stream function is also similar. This is because when the fill level increases, the amount of heat transferred to the tank also increases as the heat flux on the LH_2 side is kept uniform at 50 W/m^2.

5.5 CRYOGENIC HEAT TRANSFER CHARACTERISTICS

Generally the heat transfer processes with cryogens are very similar to those at higher temperature ranges. The superfluid helium might, however, show some differences. Also many investigations have been carried out on helium, whereas other cryogens such as LH_2 are much less investigated. The strong variation of the thermophysical properties of fluids and materials at low temperature has some impacts. The relative and absolute magnitude of the various heat transfer processes may be very different from those at room temperature and the equations become nonlinear. This has to be taken into account in the cryogenic thermal design of, e.g., thermal insulation of cryostats and transfer lines. In this section the process of heat transfer

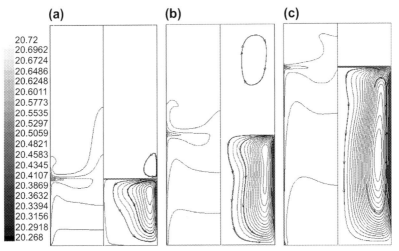

Figure 5.7 Isocontours of temperature and stream function of self-pressurization for three different fill levels at 100 s: (a) 30%, (b) 50%, and (c) 80%.

between a solid material and an adjacent cryogenic fluid is considered. The processes of interest are internal forced flow in single phase, free convection in single phase, internal two-phase flow, and pool boiling two-phase flow. The current understanding is primarily from empirical correlations based on dimensionless numbers. The issue is relevant to the design of heat exchangers, cryogenic fluid storage, superconducting magnets, and low-temperature instrumentation.

For single-phase internal flow heat transfer, commonly, the classical correlations [18] are used. This means if entrance effects are not considered, the Nusselt number is 3.656 or 4.364 depending on the boundary conditions for laminar flow, whereas for turbulent flow the Dittus—Boelter correlation is applied. For forced convection heat transfer experiments in supercritical helium, it has been found that the heat transfer coefficients are somewhat higher than those predicted by the Dittus—Boelter correlation. A slightly modified correlation has been suggested by Giarratano et al. [19,20], as it provided the best representation of many data. The correlation reads as

$$\mathrm{Nu} = 0.0259 \cdot \mathrm{Re}^{0.8} \cdot \mathrm{Pr}^{0.4} \cdot \left(\frac{T_w}{T_b}\right)^{-0.716} \tag{5.19}$$

Here, in comparison to the Dittus—Boelter correlation, the constant in front of the Reynolds number is increased from 0.023 and a temperature correction factor is introduced to take care of the temperature dependence of the thermophysical properties. T_w is the wall temperature and T_b is the fluid bulk temperature.

Heat exchangers in cryogenic systems might be of forced single-phase fluid—fluid counterflow type like a refrigerator or liquefier. LN_2 can be used for precooling of helium in a coil placed in a pool of cryogenic LN_2. Static boiling liquid—liquid might be used as a liquid subcooler in a magnet system.

For free or natural convection in low-temperature helium, correlations for the laminar and turbulent regimes have been suggested as

$$Ra < 10^9; \ Nu = 0.615 \cdot (Gr \cdot Pr)^{0.258} \tag{5.20}$$

$$Ra > 10^9; \ Nu = 0.0176 \cdot (Gr \cdot Pr)^{0.38} \tag{5.21}$$

where Gr is the Grashof number and $Gr \cdot Pr$ is the Rayleigh number, Ra. The constants and exponents in Eqs. (5.20) and (5.21) differ slightly from formulas for higher temperatures and other media [18]. In helium cryostats, strong natural convection processes with the Grashof number up to 10^{12} have been reported.

An investigation [21] on liquid and supercritical hydrogen in pool boiling has shown that the nucleate boiling heat transfer coefficient is higher at higher pressure. The critical heat flux is highest close to 0.4 MPa but is lower than the value predicted by the Kutateladze correlation for higher pressure. The reason is that the transition to film boiling is dominated not by the heat flux due to hydrodynamic instability but by the surface temperature. Another study [22] focused on pool boiling of hydrogen for normal gravity and low-gravity situations. A validated and proposed gravity scaling analysis was presented, and this might be helpful in the design process of hydrogen heat transfer systems.

General conclusions from heat transfer studies of cryogenics reveal the following:

a. Single-phase heat transfer correlations (free and forced convection) for common fluids are also applicable for cryogenic fluids, but as stated earlier, constants and exponents may differ somewhat.
b. Two-phase heat transfer in cryogenic fluids can be based on correlations for common fluids. This holds for nucleate boiling, the critical heat flux, and film boiling.
c. Transient heat transfer is governed by diffusive process for ΔT and onset of film boiling.

5.6 HYDROGEN IN AEROSPACE APPLICATIONS

A review of the scenario for using hydrogen as a fuel with a potential for zero emission was presented by Cecere et al. [23]. The history of hydrogen as propellant was summarized. The hydrogen fuel engines in low-and high-Mach-number flights were reviewed. It was concluded that hydrogen might be the fuel of the future as environmental issues and supply sources are taken into consideration. A drawback is that if hydrogen is burnt in air, NO_x is created and this might be a concern for ground transportation. For long-range aircraft transportation, use of hydrogen is believed to have some potential and the LH_2 might be stored in the fuselage to minimize the surface-to-volume ratio and prevent heat losses. Hydrogen is used in space propulsion, and it is foreseen that it will be used in future hypersonic commercial aircraft using supersonic ram jets (scramjet) because of its high energy content.

REFERENCES

[1] Glaister D, Smith J, McLean C, Mills G. Long term cryogenic storage technologies overview for NASA exploration applications. In: 42nd AIAA Thermophysics Conf, 2011—3774. Honolulu, Hawaii: AIAA; 2011. p. 2011.
[2] Hastings LJ, Plachta DW, Salerno L, Kittel P. An overview of NASA efforts on zero boiloff storage of cryogenic propellants. Cryogenics 2001;41(11—12):833—9.
[3] Chai PR, Wilhite AW. Cryogenic thermal system analysis for orbital propellant depot. Acta Astronaut 2014;102:35—46.
[4] Barsi S, Kassemi M. Investigation of tank pressurization and pressure control-part I: experimental study. ASME J Therm Sci Eng Appl 2013;5:041005.
[5] Barsi S, Kassemi M. Investigation of tank pressurization and pressure control-part II: numerical modeling. ASME J Therm Sci Eng Appl 2013;5:041006.
[6] Fu J, Sunden B, Chen X, Huang Y. Influence of phase change on self-pressurization in cryogenic tanks under microgravity. Appl Therm Eng 2015;87:225—33.
[7] Fu J, Sunden B, Chen X. Influence of wall ribs on the thermal stratification and self-pressurization in a cryogenic liquid tank. Appl Therm Eng 2014;73:1421—31.
[8] Oliveira JM, Kirk DR, Schallhorn P. Analytic model for cryogenic stratification in a rotating and reduced-gravity environment. J Spacecr Rockets 2009;46:459—65.
[9] Hirt CW, Nichols BD. Volume of fluid (VOF) method for the dynamics of free boundaries. J Comput Phys 1981;39(1):201—25. 1981.
[10] Brackbill JU, Kothe DB, Zemach C. A continuum method for modeling surface tension. J Comput Phys 1992;100:335—54.
[11] Lee WH. A pressure iteration scheme for two-phase flow modeling. In: Veziroglu TN, editor. Multiphase transport fundamentals, reactor safety, applications, 1. Washington, DC: Hemisphere Publishing; 1980.
[12] Liu Z, Sunden B, Yuan J. VOF modeling and analysis of filmwise condensation between vertical parallel plates. Heat Transf Res 2012;43(1):47—68.
[13] Fu J, Sunden B, Chen X. Analysis of self-pressurization phenomenon in a cryogenic fluid storage tank with VOF method, 2013;ASME IMECE2013—63209 (paper no).

[14] Kumar SP, Prasad BVSSS, Venkatarathnam G, Ramamurthi K, Murthy SS. Influence of surface evaporation on stratification in liquid hydrogen tanks of different aspect ratios. Int J Hydrogen Energy 2007;32:1954—60.

[15] Seo M, Jeong S. Analysis of self-pressurization phenomenon of cryogenic fluid storage tank with thermal diffusion model. Cryogenics 2010;50:549—55.

[16] Gursu S, Sherif SA, Veziroglu TN, Sheffield JW. Analysis and optimization of thermal stratification and self-pressurization effects in liquid hydrogen storage systems, part 1: model development. ASME J Energy Resour Technol 1993;115:221—7.

[17] Das SP, Chakraborty S, Dutta P. Studies on thermal stratification phenomenon in LH2 storage vessel. Heat Transf Eng 2004;25:54—66.

[18] Sunden B. Introduction to heat transfer. UK: WIT Press; 2012.

[19] Giarratano PJ, Arp VD, Smith RV. Forced convection heat transfer to supercritical helium. Cryogenics 1971;11:385—93.

[20] Giarratano PJ, Jones MC. Deterioration of heat transfer to supercritical helium at 2.5 atmospheres. Int J Heat Mass Transf 1975;18:649—53.

[21] Tatsumoto H, Shirai Y, Shiotsu M, Naruo Y, Kobayashi H, Inatani Y. Heat transfer characteristics of horizontal wire in pools of liquid and supercritical hydrogen. J Supercond Nov Magnetism 2015;28(4):1185—8.

[22] Wang L, Zhu K, Xie F, Ma Y, Li Y. Prediction of pool boiling heat transfer for hydrogen in microgravity. Int J Heat Mass Transf 2016;94:465—73.

[23] Cecere D, Giacomazzi E, Ingenito A. A review on hydrogen industrial aerospace applications. Int J Hydrogen Energy 2014;39:10731—47.

CHAPTER 6

Aerospace Heat Exchangers

6.1 INTRODUCTION

Heat exchangers are equipment being used for transfer of heat between two or more fluids at different temperatures [1]. They are widely used in diverse energy and process applications such as power plants, automotives, space heating, refrigeration and air-conditioning systems, aerospace industry, petrochemical processes, electronics cooling, and environment engineering. Specifically in the aerospace industry, heat exchangers are mainly used in three systems: (1) gas turbine cycle, (2) environmental control system (ECS), and (3) thermal management of power electronics.

Heat exchangers may be classified in various ways, e.g., based on transfer processes, surface compactness, construction features, flow arrangements, and heat transfer mechanisms. Design and sizing of heat exchangers are generally complicated. The basic theory and methodology for heat exchanger design are detailed in, e.g., Shah and Sekulić [2], Sundén [1], and Wang et al. [3]. In general, heat transfer, pressure loss, cost, materials, operating limits, size, and weight are important parameters. For aerospace industry, in particular, the weight and the size of the heat exchanger are most important.

Various enhancement techniques (e.g., extended surfaces such as the fins) are adopted in heat exchangers to augment the heat transfer surface area or the heat transfer coefficient, or both. The goal of enhancement techniques might be to reduce the size of the heat exchanger for a given duty, to increase the capacity of an existing heat exchanger, or to reduce the approach temperature difference [4]. Because of the advancement in manufacturing processes, enhanced geometries have been gradually used in nearly all heat exchangers. Heat transfer enhancement techniques can be classified as either passive (no external power needed) or active (external power required). Passive techniques employ special surface geometries (e.g., surface coatings, fins, inserts, curved tubing, surface tension devices) or additives (e.g., microparticles, nanoparticles, phase change materials, surfactants) for heat transfer enhancement. Bergles [5] defined the four generations of heat transfer technology using passive techniques. Most of

Heat Transfer in Aerospace Applications
ISBN 978-0-12-809760-1
http://dx.doi.org/10.1016/B978-0-12-809760-1.00006-5

the heat transfer enhancement techniques were covered by Webb and Kim [6]. Implementation of heat transfer enhancement techniques in aerospace heat exchangers might help reduce fuel consumption and the size of the heat exchanger. Passive enhancement techniques such as enhanced surface geometries are preferred, especially surface enhancement elements that largely improve heat transfer with a comparably low pressure drop penalty. It is important to note that designers need to evaluate the durability of the enhanced geometries and their effects on the structural integrity of the aerospace heat exchangers when considering the use of enhanced geometries.

6.2 APPLICATIONS OF AEROSPACE HEAT EXCHANGERS

6.2.1 Gas Turbine Cycles

Heat exchangers can be implemented in gas turbine cycles to increase thermal efficiency and to reduce pollutant and greenhouse gas emissions. According to the ACARE (Advisory Council for Aviation Research and Innovation in Europe) Flightpath 2050, aviation industry should allow a 75% reduction in CO_2 emissions per passenger kilometer, a 90% reduction in NO_x emissions, and a 65% reduction in noise emissions relative to the capabilities in 2000 [7]. For a conventional gas turbine cycle, the thermal efficiency mainly depends on the overall pressure ratio and the turbine inlet temperature. The overall pressure ratio indicates the ratio of the compressor outlet pressure to the compressor inlet pressure. The overall pressure ratio and the turbine inlet temperature are limited by the maximum pressure and temperature that the turbine blades can withstand. New materials and innovative cooling concepts for the critical components can further increase the overall pressure ratio and the turbine inlet temperature. One technique to improve the overall pressure ratio for a given compression work is to introduce multistage compression with intercooling, in which the gas is compressed in stages and cooled between each stage by passing the gas through a heat exchanger called an intercooler. For aero gas turbine engines with relatively high overall pressure ratios, the compressor is split into a low pressure compressor (LPC) and a high pressure compressor (HPC). Therefore, an intercooler can be introduced between the LPC and the HPC. At the outlet of the LPC, the compressed gas has a relatively higher temperature. Using a crossflow or counterflow air-to-air heat exchanger, with compressed air flowing on one side and low-temperature ram air flowing on the other side, the compressed air can be cooled before entering the HPC. The steady-flow compression work or the pressure ratio for a given compression work is proportional to the specific volume of the compressed air [8]. The intercooler decreases the temperature and hence decreases the specific volume of the compressed air, which improves the thermodynamic cycle efficiency.

In gas turbine engines, the temperature of the exhaust gases leaving the turbine is often considerably higher than that of the air leaving the HPC. A regenerator or a recuperator, i.e., a crossflow or counterflow heat exchanger, can be incorporated to transfer heat from the hot exhaust gases to the compressed air. Accordingly, the thermal efficiency increases because the portion of the energy of the exhaust gases that is supposed to be rejected to the surroundings is recovered to preheat the air entering the combustion chamber. The recuperator is more advantageous when an intercooler is used because a greater potential for recuperation exists. For very high overall pressure ratios, e.g., greater than 50, when the temperature of the hot exhaust gases is only slightly higher than that of the compressed air, a recuperator is not effective, especially considering its cost, size, and weight. Fig. 6.1 shows a conceptual sketch that compares the thermal efficiencies of different gas turbine cycles with the overall pressure ratios. In general, the intercooled and recuperated gas turbine cycle is effective at relatively low overall pressure ratios, e.g., less than 30. The intercooled gas turbine cycle

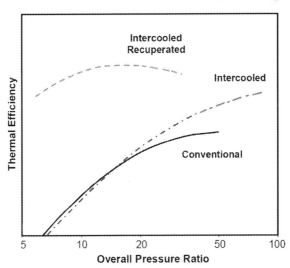

Figure 6.1 Thermal efficiency versus overall pressure ratio for different gas turbine cycles. *(Adapted from Sieber J, Overview NEWAC (new aero engine core concepts); April 2009.)*

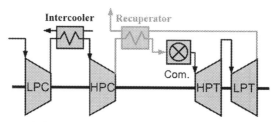

Figure 6.2 Intercooled and recuperated gas turbine cycle. *LPC,* low-pressure compressor; *HPC,* high-pressure compressor; *Com,* compressor; *HPT,* high-pressure turbine; *LPT,* low-pressure turbine *(Adapted from Sieber J, Overview NEWAC (new aero engine core concepts); April 2009.)*

without recuperation is only effective at very high overall pressure ratios. Fig. 6.2 illustrates an intercooled and recuperated gas turbine cycle.

In addition, precoolers, such as cryogenic fuel-cooled air heat exchangers, that immediately downstream the air intake to precool the air entering the engine can be used for high-speed jet engines to save fuel consumption for a given overall pressure ratio or to increase the overall pressure ratio for a given compression work.

Besides, modern gas turbine engines operate at turbine inlet air temperature levels that are beyond the material maximum temperature. Hence, hot engine components such as the turbine blades must be cooled to assure their structural integrity. Typically the turbine blades are cooled by bleed air from the compressor, which, although cooler than the turbine, has been already heated up by the work done on it by the compressor [8]. Engine bleed reduces thrust and increases fuel burn. The effect is a function of the mass flow rate of the bleed air. If a heat exchanger is used to cool the bleed air before its introduction for blade cooling, the amount of bleed air required by the blade cooling can be reduced, which results in an improved engine performance with a consequent reduction in specific fuel consumption (the rate of fuel consumption per unit of power output). Such a heat exchanger might be a fuel-to-air heat exchanger. The relatively hot bleed air can be cooled by the cool engine fuel. On one hand, the cooling capacity of the bleed air increases as its temperature is decreased. On the other hand, the energy extracted by the fuel is reintroduced into the propulsive cycle as the heated fuel is burned in the combustor.

6.2.2 Environmental Control System

The ECS is employed in aerospace vehicles such as large commercial aircraft to provide comfortable flight conditions for passengers.

There are two kinds of air-conditioning systems on an aircraft: air cycle air-conditioning and vapor cycle refrigeration system. For air cycle air-conditioning, the source of air used by the ECS is typically bleed air from the gas turbine compressor, with a relatively high pressure and temperature. The hot bleed air must first be cooled to an acceptable temperature before entering into the passenger cabin. This is performed using an air-conditioning package that is composed of several units, including a number of heat exchangers cooled by ambient ram air. As shown in Fig. 6.3 the hot bleed air from the engine compressor is metered through a bleed air valve and then cooled by the primary heat exchanger. Then the air passes to the air cycle machine for pressure adjustment and finally into the secondary heat exchanger for temperature adjustment before entering the cabin. The primary and secondary heat exchangers shown in Fig. 6.3 are air-to-air heat exchangers.

A vapor cycle refrigeration system, in which the refrigerant undergoes phase changes, is basically the same refrigeration cycle that is used in home air conditioners. The vapor cycle refrigeration system is a closed system used for transferring heat from inside the cabin to outside the cabin. Heat exchangers such as evaporators and condensers are important devices in the refrigeration cycle. Specifically the liquid refrigerant decreases its pressure by passing through a thermal expansion valve before entering the

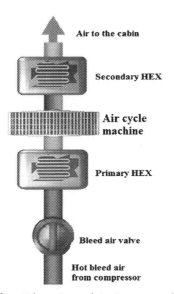

Figure 6.3 A schematic of an Airbus air-conditioning system [10]. *HEX*, heat exchanger.

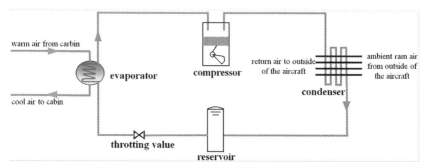

Figure 6.4 A vapor cycle refrigeration system.

evaporator. The liquid refrigerant in the evaporator changes into vapor, absorbing the heat energy from the cabin air. The vapor is then compressed and becomes hot. The hot vapor refrigerant transfers its heat energy to the outside ram air in a condenser and thus the refrigerant condenses back into liquid to repeat the cycle. Therefore, the warm cabin air is constantly replaced or mixed with cool air in the aircraft cabin to maintain a comfortable temperature. An example of a vapor cycle refrigeration system is shown in Fig. 6.4.

6.2.3 Thermal Management

The trend to pack current and future aerospace and military platforms with power-hungry, heat-generating electronic components and systems drives the need for efficient, effective, compact, and lightweight thermal management systems. Shrinking electronics packaging and high-density integration of power electronics to enable more power and functionality in a small unit, coupled with the extremes of military and aerospace environments, constantly place ever-increasing demands on precise and smart thermal management solutions to maintain junction temperatures below levels that degrade performance. The temperature of the components can be maintained at a safe level by air cooling for low-heat-flux components. As the heat flux increases, the limits of air cooling technology are being approached, i.e., forced air heat sinks have become significantly larger, more expensive, and more complex. Liquid cooling is promising to provide the needed level of thermal performance, with an increase in energy efficiency compared to traditional air cooling. The liquid must pass through the heat sources to carry away the heat, resulting in a temperature increase in the liquid. Then the liquid passes through a liquid-

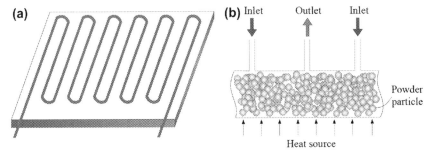

Figure 6.5 Cold plate examples: (a) a tube liquid cold plate and (b) a powdered metal cold plate.

to-air heat exchanger, such as a plate-fin heat exchanger (PFHE), to transfer the stored heat to the air. An important component for liquid cooling is the cold plate. The cold plate must be able to dissipate the waste heat efficiently within a relatively small unit. Fig. 6.5 shows examples of cold plates, namely, the tube liquid cold plate and the powdered metal cold plate [11], with the former suitable for low heat fluxes and the latter for high heat fluxes. Cold plates with embedded microchannels are promising to dissipate high heat fluxes. In many cases, heat pipes are embedded in cold plates to increase the effective thermal conductivity of the cold plates. Basically the cold plates should be designed by conforming to the heat-generating components. Besides, the cold plates should be optimized to augment the thermal performance while maintaining a relatively low pressure drop penalty by properly designing, e.g., the liquid flow passages and manifold distributions inside the cold plates. A liquid cooling concept is sketched in Fig. 6.6.

For ultrahigh heat flux components such as converters or densely packed computing devices with heat fluxes greater than 100 W/cm^2, liquid cooling

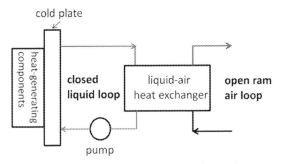

Figure 6.6 A concept of a liquid cooling solution.

might not be sufficient enough to cool the heat-generating components with certain form factors. Besides, similar to air cooling, liquid cooling cannot provide relatively uniform cooling as the liquid temperature increases along the cooling path. One way of providing an efficient cooling system for high-power-density electronic devices is by using an evaporative cooling circuit. This circuit brings a liquid into thermal contact with the heat-generating components via an evaporating unit to effectively remove heat by utilizing the latent heat of the working fluids and at the same time to maintain a relatively uniform temperature distribution. Compared to air cooling and single-phase liquid cooling, microchannel evaporative cooling has a low pumping power requirement relative to the quantity of heat to be removed [12]. At the same total flow rates, the pressure drop in evaporative micro-channels is much higher than that for single-phase liquid flow in conventional channels. However, to dissipate the same amount of heat, much less liquid is required in microchannel evaporative cooling than that in single-phase liquid cooling. Fig. 6.7 shows an evaporative cooling loop with a pump, a cold plate evaporator, an air-cooled condenser, and a reservoir. It has been stated that such an evaporative cooling system with refrigerant R134a as the working fluid can provide twice the cooling capacity in half the size or in a size less than that of air- or water-cooling systems [13].

Besides, the temperature of lubricating oil needs to be controlled. The excessive heat caused by friction in devices such as gearboxes and compressors increases the lubricant oil temperature. As the oil viscosity decreases with increment in temperature, the lubricant oil film becomes thinner and collapses, which in turn causes more friction and heat. Fuel or ambient ram air can be used as working fluids in fuel-cooled oil cooler or air-cooled oil cooler to cool the lubricants.

To conclude, typical applications of aerospace heat exchangers are summarized in Fig. 6.8.

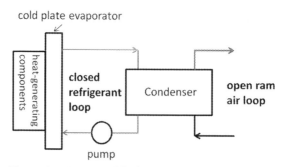

Figure 6.7 An exemplified evaporative cooling system.

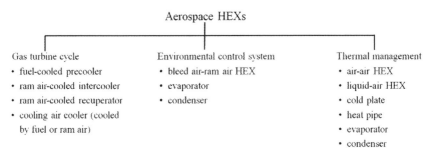

Figure 6.8 Applications of aerospace heat exchangers (HEXs).

6.3 GENERAL DESIGN CONSIDERATIONS FOR AEROSPACE HEAT EXCHANGERS

The typical features of aerospace heat exchangers are briefly listed below.

• Compactness and lightweight: Compact heat exchangers are promising because of the limited space available for heat transfer in aerospace industry. Light materials such as aluminum alloy and metal foam might be preferred if the operating pressure and temperature are less than the maximum allowable pressure and temperature of the materials.

• Relatively high temperature and pressure for some aerospace heat exchangers such as recuperators: Compressed air leaving the HPC flows on one side of the recuperator and the exhaust gas exits from the turbine flows on the other side. For example, for an overall pressure ratio of 50, the compressed air leaving the HPC has a pressure of 50 bar and a temperature up to 1000 K. The materials for high-temperature heat exchangers should be able to maintain structural integrity to cope with large temperature differences.

• High effectiveness and minimum pressure loss: The heat-exchanging surfaces and flow paths should be designed in such a way to give a high thermal conductance with a minimum pressure loss. Besides, the bypass duct and the inlet and outlet ducts to and from the heat exchangers need to be arranged properly to minimize momentum pressure losses.

Compact heat exchangers such as PFHEs, printed circuit heat exchangers (PCHEs), micro heat exchangers, primary surface heat exchangers, heat pipe heat exchangers, and various heat exchangers using new materials such as ceramics and foam materials have been used or are promising for use in the aerospace industry.

6.4 PLATE-FIN HEAT EXCHANGERS

A PFHE consists of flat plates and finned chambers to transfer heat between fluids. PFHE is often categorized as a compact heat exchanger with a relatively high heat transfer surface area to volume ratio. Fig. 6.9 shows an example of a PFHE. In aerospace industry, aluminum alloy PFHEs have been used for 50 years for their compactness and lightweight properties. Most commonly this heat exchanger type is for gas-to-gas heat exchange. Various fin geometries such as triangular, rectangular, wavy, louvered, perforated, serrated, or the so-called offset strip fins are used to separate the plates and create the flow channels [6]. Several fin geometries are shown in Fig. 6.10.

The common fin thickness ranges from 0.046 to 0.20 mm and the fin height ranges from 2 to 20 mm. Manufacturing technologies nowadays can fabricate PFHEs with thin fins and a large fin density, without compromising the durability. The material mass for fins can be minimized by reducing the fin thickness as permitted by the design and manufacturing considerations and by increasing the fin density to maintain the desired fin heat transfer. This is primarily because fins are used for increasing the surface area for convection heat transfer. From a heat transfer perspective, the fin thickness is chosen the thinnest and is of secondary importance for industrial heat exchangers because of the limited range of permissible fin sizes [2]. The lowest fin thickness is primarily restricted by the mechanical strength of the fin material. In general, fins require one-third to one-fourth surface thickness to cost-effectively utilize the material. Thus modern fin designs have the highest possible fin density and lowest possible fin thickness permitted by design to make the most use of the fin material in a heat exchanger. This trend will continue with improvements in material and manufacturing technologies.

Figure 6.9 A crossflow PFHE.

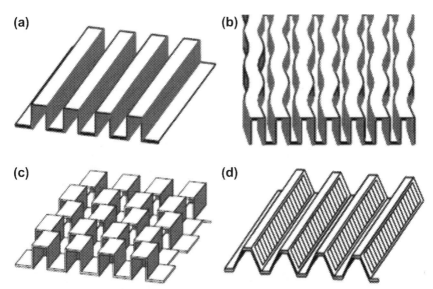

Figure 6.10 Types of enhanced fin geometries: (a) rectangular fin, (b) wavy fin, (c) offset strip fin, and (d) louvered fin.

On the other hand, fin height is limited by the structural integrity, as shorter fins have much higher column structural strength. Besides, shorter thin fins have a high efficiency and can reduce conjugate heat transfer, i.e., the transfer of heat from higher temperature zones to the lower temperature zones, through the solid fins by heat conduction. The fin efficiency is maintained at 90% but is higher in most PFHEs, particularly in the aerospace industry. Hence, a shorter fin is more desirable from both the heat transfer and mechanical strength aspects.

There are manufacturing difficulties associated with the true counter-flow arrangement for PFHEs concerning complex inlet and outlet header designs. Crossflow arrangement is common for PFHEs in the aerospace industry for air-to-air heat transfer applications. Besides, fins are not densely spaced to allow for desired fluid flow rates with allowable pressure losses and to have minimum fouling potential. Another disadvantage of common PFHEs is that they can only handle relatively low pressures (less than 10 bar) and low temperatures depending on the materials used for fabrication. Recently, diffusion bonding has been introduced to fabricate PFHEs to handle high temperatures and pressures.

For aerospace applications, at least one of the fluids in the PFHEs is a gas, e.g., ambient ram air or bleed air. For heat transfer between two gases, the finned surfaces in PFHEs will provide a substantial size and weight reduction of the heat exchanger. In PFHEs, the flow regime is likely to be laminar flow because of the small hydraulic diameter and low density of gases. The finned surfaces in PFHEs are designed to promote fluid mixing and boundary layer destruction, which are very effective for low-Reynolds (Re)-number laminar flows. For heat transfer between a gas and a liquid, the thermal resistance of the gas side is much higher than that of the liquid side. Hence finned surfaces are primarily used in the gas side to enhance heat transfer coefficient and surface area of the gas side. For aerospace heat exchangers, pressure drop requirements are often more difficult to meet than heat transfer requirements [14]. Finned passages are known to exhibit a greater increase in pressure drop than heat transfer. The key to PFHE design is to reduce the flow length such that the overall pressure drop requirement can be met, even though the pressure drop per unit length is high.

More information about PFHEs and the effects of various fin geometries on heat transfer and pressure drop of PFHEs are available in Ref. [6].

6.5 PRINTED CIRCUIT HEAT EXCHANGERS

The concept of PCHEs was first invented in the early 1980s at the University of Sydney in Australia. It has been commercially manufactured by Heatric Ltd in the United Kingdom since 1985. A brief introduction of PCHE is given in Reay et al. [4]. The PCHE derives its name from the procedure used to manufacture the plates that form the core of the heat exchanger; the fluid flow passages are produced by chemical milling, a technique similar to that used to manufacture printed circuit boards in the electronics industry. A benefit of this design is the flexibility it affords in terms of flow passage geometry. The fluid flow passages for PCHE have approximately a semicircular cross section. Channel depths and channel widths are typically in the range 0.5–2.0 mm and 0.5–5.0 mm, respectively, and may be larger for some streams. PCHEs are highly compact because of their high surface area-to-volume ratios ($>2500 \text{ m}^2/\text{m}^3$) [15]. A PCHE is a high-efficiency plate-type compact heat exchanger, which is typically four to six times smaller and lighter than a shell-and-tube heat exchanger of equivalent performance. Fig. 6.11 shows an example of a PCHE.

The typical manufacturing process of PCHEs involves photochemical etching and diffusion bonding. Photochemical etching uses strong chemical

Figure 6.11 A diffusion-bonded PCHE [16].

etchants to remove unwanted workpiece material by controlled dissolution. This process involves corrosive oxidation of the selected areas of the base plate and does not alter the internal structure of the material and the material properties such as grain structure, hardness, and ductility. Fine grooves are photochemically etched on one side of the base plate, forming the flow passages. The etched-out plates are thereafter alternatively joined by diffusion bonding, which results in compact, extremely strong, all-metal heat exchanger cores. Diffusion bonding is a solid-state joining process by which two nominally flat surfaces are joined at an elevated temperature using an applied interfacial pressure for a period ranging from a few minutes to a few hours [17]. The success or failure of diffusion bonding is primarily controlled by three variables, i.e., the bonding temperature, the bonding pressure, and the holding time. Furthermore the bonding surfaces should have a good surface finish and should be clean and free from oxide films and adsorbed grease. With carefully controlled bonding variables, this process can offer parent metal strength and high-pressure containment capability. The complete heat exchanger core is made by welding together as many of these blocks as the thermal duty of the heat exchanger requires. Finally, fluid headers and nozzles are welded to the cores. The PCHE is capable of being operated at both high temperature and high pressure, with a relatively thin wall. It is capable of operating at temperatures within the range of 73–1173 K and can cope with pressures up to 600 bar. PCHEs are very flexible in terms of the variety of fluid types and flow configurations. In addition to the high efficiency and compactness, wide operating range, and flexibility, the great potential of PCHE is also illustrated by its safety features, such as the very low risk of leakage.

A concern about PCHEs in the aerospace industry is the relatively large pressure drop, which is roughly inversely proportional to the channel hydraulic diameter. Each flow channel in a PCHE plate can be regarded as a small pipe with many bends; thereby, the pressure drop in the channels is relatively high because of the reduction in hydraulic diameter, the longer flow path of wavy channels, and the flow separation at channel bending points. Therefore, a PCHE design that can reduce pressure drop in channels is promising in view of the cycle efficiency. The flow paths can be tailored to improve the overall performance of PCHEs. For example, flow passage designs are developed to reduce the pressure drop compared to the Heatric zigzag pattern. Tsuzuki et al. [18] adopted discontinuous S-shaped fins and achieved a pressure drop at only 20% of the conventional zigzag passage, as shown in Fig. 6.12.

Streamlined fin shape airfoil was adopted in a PCHE model in the 3D (three-dimensional) numerical analysis by Kim et al. [19], as shown in Fig. 6.13. Simulation results showed that in the airfoil shape fin PCHE, total heat transfer rate per unit volume was almost the same as for the zigzag channel PCHE and the pressure drop was reduced to one-twentieth of that

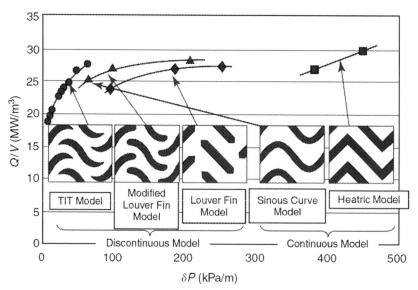

Figure 6.12 Comparison of five flow passage patterns on PCHEs, Q/V is the heat transfer rate per unit volume [18]. *TIT*, turbine inlet temperature.

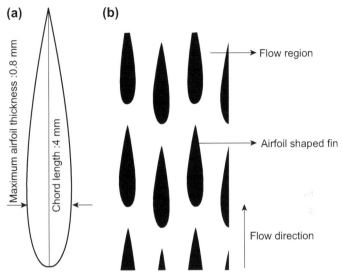

Figure 6.13 (a) The National Advisory Committee for Aeronautics (NACA) 0020 airfoil shape and (b) channel configuration of airfoil fin PCHE [19].

in the zigzag channel PCHE. In the airfoil shape PCHE, the enhancement of heat transfer area and the uniform flow configuration contributed to achieve the same heat transfer performance as for the zigzag channel PCHE. The pressure drop reduction in the airfoil shape fin PCHE was caused by suppressing the generation of separated flow due to the streamlined shape of airfoil fins.

6.6 MICRO HEAT EXCHANGERS

Micro- and miniscale heat exchangers (μHEXs) are heat exchangers in which at least one fluid flows in lateral confinements such as channels or small cavities with dimensions less than around 1 mm in size. Some PFHEs and PCHEs can also be categorized as μHEXs. Typically, the fluid flows through a cavity that is called a microchannel. This process intensification technology exploits heat transfer enhancement resulting from structurally constrained fluid streams in microchannels, which reduces the thermal resistance considerably. Micro heat exchangers have been demonstrated with high heat transfer coefficients, approximately one order of magnitude higher than the typical values of conventional heat exchangers. Therefore, as a typical miniaturized process device, micro heat exchangers have attracted widespread applications because of their high thermal performance, compactness, small size, and lightweight [20], with the aim of process intensification. Since the pioneering work of Tuckerman and Pease [21] in 1981, which exhibited for the first time a high heat-flux dissipation of up to 790 W/cm^2 by using microchannels, a lot of effort have been devoted to investigate single-phase and two-phase fluid flow and heat transfer characteristics in μHEXs [22].

Micro heat exchangers are used in diverse energy and process applications such as electronics cooling, automotive and aerospace industries, chemical process intensification, refrigeration and cryogenic systems, and fuel cells. Micro heat exchangers have many advantages over conventional heat exchangers. First, in the scaling down from macro- to microscale, the volume decreases with the third power of the characteristic linear dimensions, whereas the surface area only decreases with the second power. Therefore, micro heat exchangers have relatively larger surface area-to-volume ratios that enable higher heat transfer rates than conventional heat exchangers. Second, fast fluid acceleration and close proximity of the bulk fluid to the wall surface in μHEXs give high heat transfer coefficients. For single-phase fully developed internal laminar flow, a constant Nusselt number (Nu = hd$_h$/k) implies that the single-phase heat transfer coefficient increases as the hydraulic diameter decreases. In addition, compactness and high heat-flux dissipation are required as the scale of the devices decreases, whereas the power density increases. Micro heat exchangers are expected to manage the high heat-flux dissipation for miniaturized devices to guarantee the reliability and safety during operation.

Although the development of μHEXs is mainly driven by demands from industrial process intensification (with reduced inventories, reduced footprints, and reduced weight) and sustainable environment, progress in the area of material science has been attributed to the manufacturers of μHEXs. Both traditional and modern micromachining techniques can be applied to fabricate micro heat exchangers [23]. Traditional techniques such as computer numerical control (CNC) mill, micro sawing, dicing, and micro deformation are possible choices. Modern fabrication techniques include laser micromachining, deep reactive ion etching (DRIE), LIGA [Lithographie, Galvanoformung, Abformung (lithography, electroplating, and molding)], wafer bonding, diffusion bonding, electrodischarge machining, and 3D printing. Advantages and disadvantages of several micromachining methods are given in Table 6.1. Fig. 6.14 shows several examples of fabricated μHEXs.

Table 6.1 Several Micromachining Methods

	Micro Deformation	Laser Micromachining	DRIE	LIGA
Geometries	Rectangular	Unlimited	Rectangular, circular, trapezoidal, etc.	Rectangular
Materials	Metal and nonmetal	Metal and glass	Metal, silicon, glass, and PDMS	Metal and silicon
Channel range	250 channels/inch	Scale: nanometer to millimeter	Scale: nanometer to millimeter	Tens of micrometers to several millimeters
Advantages	Low cost, fast	Low manufacturing uncertainty	Low to high aspect ratios	High aspect ratio, vertical smooth side walls
Disadvantages	Some materials require posttreatment	Too expensive	Slow process (1 day)	Expensive

From Sundén B, Wu Z, Advanced heat exchangers for clean and sustainable technology. In: Handbook of clean energy systems, editors J. Yan, John Wiley & Sons.

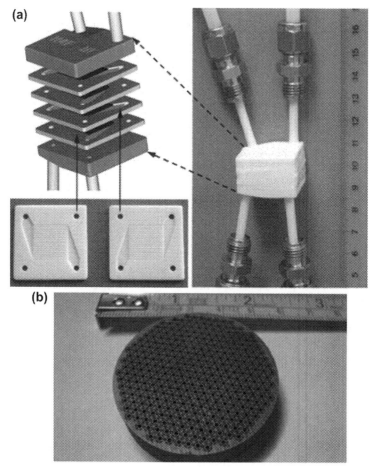

Figure 6.14 Examples of micro heat exchangers: (a) ceramic micro heat exchanger [24] and (b) 3D-printed heat exchanger by Sustainable Engine Systems Ltd. [4].

For single-phase air and liquid flow in μHEXs with channel hydraulic diameters ranging from about 150 μm to 1 mm, conventional theories and correlations are still able to predict the flow and heat transfer characteristics in microchannels when the following effects are considered appropriately: (1) microchannel geometry, (2) entry and exit effects, (3) surface roughness, (4) developing flow, (4) conjugate heat transfer, (5) viscous dissipation, and (6) temperature-dependent properties. For hydraulic diameters less than about 100 μm, more scaling effects, such as electric double layer (EDL) effects, rarefaction, and compressibility effects should be considered [25]. The above-mentioned possible scaling effects, often negligible in conventional channels, may now have a significant influence for micro-channels. Conjugate heat transfer is accordingly discussed, as it is important for designing micro heat exchangers, PFHEs, and PCHEs, all relevant for aerospace applications.

Conjugate heat transfer refers to the ability to conduct heat through solids, coupled with convective heat transfer in a fluid with coupled boundary conditions. Because of the temperature differences within the solid wall structure, heat will flow through the solid walls from the warmer end to the cooler end, flattening the solid temperature distribution. The temperature difference between the two fluids decreases and therefore, the potential to transfer heat also decreases. For a μHEX with parallel micro-channels of relatively short flow lengths, 3D effects in the solid walls and in the fluid might significantly modify the heat transfer behavior. Maranzana et al. [26] introduced a dimensionless number (M) to evaluate if conjugate heat transfer should be considered

$$M = \frac{\text{axial heat conduction in the wall}}{\text{convective heat transfer in the fluid}} = \frac{k_s}{\rho_f c_{pf} u} \frac{A_s}{A_f L}$$

$$= \frac{k_s}{k_f} \frac{A_s}{A_f} \frac{d_h}{L} \frac{1}{RePr} \qquad (6.1)$$

where A_s and A_f indicate the cross-sectional areas of the solid part and the fluid channels, respectively. If $M < 0.01$, the axial heat conduction in the solid walls can be neglected. Fig. 6.15 shows the variation of M with

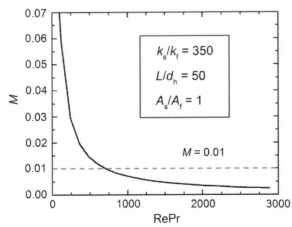

Figure 6.15 The variation of M with RePr for a typical microchannel heat sink. The *red dashed line* indicates M = 0.01. When M > 0.01, the conjugate heat transfer should be considered in the numerical modeling [26].

RePr (product of Reynolds and Prandtl numbers) for a typical microchannel heat sink with a thermal conductivity ratio (k_s/k_f) of 350, a length-to-diameter ratio (L/d_h) of 50, and a cross-sectional area ratio (A_s/A_f) of 1. As shown in Fig. 6.15, M becomes greater than 0.01 when RePr is less than around 700. Thus conjugate heat transfer needs to be considered in heat exchanger design at relatively low Re numbers ($\sim$100 to $\sim$700 depending on the Pr value) in laminar flow. Different from microchannel heat sinks, for conventional channels the wall thickness is comparably much smaller than the channel diameter. Therefore for conventional channels the cross-sectional area ratio (A_s/A_f) is much less than 1 and the M value is less than 0.01 even at very low Re values.

Recently, Honeywell Aerospace has developed microchannel heat exchangers for aerospace use. It is stated that these microchannel heat exchangers may offer significant decreases in volume and/or weight compared to the state-of-the-art aircraft PFHEs using offset plate and fin surfaces [14]. Depending on design, microchannel heat exchangers can be used for various types of units such as air-to-air ECS or thermal management of power electronics and lubricating oils. For example, cold plates embedded with microchannels are promising for use in electronics cooling. Microstructured heat exchangers can be used as fuel-oil heat exchangers, evaporators, condensers, etc. Aluminum and higher temperature materials such as alloys and ceramics can be used to fabricate microchannel heat exchangers. It was reported by Strumpf and Mirza [14] that a microchannel heat exchanger offered a 42% reduction in core size compared to an existing PFHE.

Nacke et al. [27] proposed crossflow microchannel heat exchangers intended for high-Mach-aircraft gas turbine engines. The compressor air that is used to cool turbine blades and other components in a high-Mach-number engine is too hot, so the aircraft fuel is used to precool the compressor air in a crossflow microchannel precooler. The precooler consists of a large number of miniature closely spaced modules. Within each module, the fuel flows through a series of parallel microchannels, whereas the air flows externally over rows of short straight fins perpendicular to the direction of fuel flow. The crossflow microchannel air-to-fuel precooler is illustrated in Fig. 6.16.

The overall thermal performance of μHEXs can be further enhanced effectively by tailoring the flow paths to improve heat transfer with a small pressure drop penalty [22]. However, the structural integrity should always be maintained when incorporating enhancement elements.

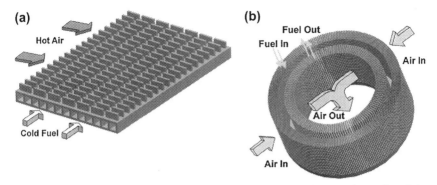

(a)

Hot Air

Cold Fuel

(b)

Fuel Out
Fuel In

Air In

Air Out

Air In

Figure 6.16 (a) Crossflow microchannel precooler module and (b) stacking of modules in a radial inflow precooler design [27].

6.7 OTHER AEROSPACE HEAT EXCHANGERS

6.7.1 Primary Surface Heat Exchangers

Primary surface heat exchangers refer to heat exchangers wherein substantially all the material that conduct heat between two media comprises the walls separating the two media. This type of heat exchangers primarily consists of plates or sheets and have no separate or additional internal members, such as fins, so that the exchanger is constructed of plates or sheets, each side of which is in contact with a different fluid, and heat transfer occurs substantially and directly between the plates and the fluid. In contrast, extended surface heat exchangers, or secondary surface heat exchangers, contain a substantial amount of material in the form of fins, which do not separate the media. The PFHE and micro heat exchanger are mostly extended surface heat exchangers. The main attributes of primary surface heat exchangers are that the surface geometry is 100% effective (no fins) and sealing can be accomplished by welding without the need for an expensive and time-consuming high-temperature furnace brazing operation. Primary surface heat exchangers can be used as intercoolers and recuperators in the aerospace industry. Fig. 6.17 shows a part of a cross-corrugated primary surface heat exchanger manufactured by selective laser melting by Rolls-Royce for an intercooled core [28].

6.7.2 Heat Pipe Heat Exchanger

Heat pipes of various capillary wick structures are attractive in the area of spacecraft cooling and temperature stabilization because of their lightweight, zero maintenance, and reliability. A heat pipe consists of an

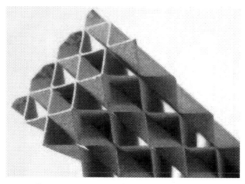

Figure 6.17 A part of a cross-corrugated primary surface heat exchanger manufactured by selective laser melting by Rolls-Royce [28].

evaporation zone, an adiabatic zone, and a condensation zone. Liquid in contact with the evaporation zone turns into vapor by absorbing heat, and vapor then travels along the heat pipe to the condensation zone and condenses back into liquid, releasing the latent heat. The liquid then returns to the evaporation zone through capillary forces or other external forces. Loop heat pipes and micro/miniature heat pipes are promising types suitable for cooling solutions. Micro heat pipes are heat pipes with a height less than about 1 mm. Micro and miniature heat pipes are able to dissipate high heat fluxes up to 100 W/cm^2 [29]. A major advantage of heat pipes is that no external power is required. For high heat fluxes, heat pipes need a relatively larger area for heat dissipation from the heat pipe condensation zone to the environment, which might be a concern because of the limited space in aerospace. Fig. 6.18 shows a sketch of a conceptual flat micro heat

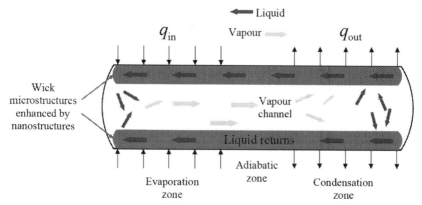

Figure 6.18 A flat micro heat pipe (not to scale).

pipe. The depth of the micro heat pipe is less than 1.0 mm. The width and the length of the evaporation zone depend on the form factor of the electronic components. The length of the condensation zone can be designed by the required dissipated heat and the cooling mode of the external heat sink.

6.7.3 Heat Exchangers Using New Materials

6.7.3.1 Foam Materials

Light materials are preferred in the manufacturing of aerospace heat exchangers to save fuel consumption. Aluminum alloy heat exchangers such as aluminum alloy PFHEs are common in the aerospace industry. Foam materials such as open-cell porous metallic foams and graphite foams have received increased attention for use in commercial heat exchangers because of their lightweight, improved thermal performance, high compactness, attractive flexibility to be formed in complex shapes, good stiffness/strength properties, and low cost via the metal sintering route for mass production, as well as because some materials can be used at high temperatures up to 1200 K. Because of the interconnection of pores and tortuosity of open-cell foams, fluid mixing and convection heat transfer are greatly enhanced. The key parameters that influence the thermohydraulic performance of foams are porosity, pore density (pores per inch), mean pore diameter, surface area-to-volume ratio, effective thermal conductivity, and permeability. The pertinent correlations for flow and thermal transport in metal foam heat exchangers were categorized in Ref. [30]. Fig. 6.19 shows open-cell metallic foam and a metal foam heat exchanger. Graphite foam is attractive, as the density of graphite foam ranges from 200 to 600 kg/m^3, which is about 20% of that of aluminum. However, the tensile strength of graphite foam is much less than that of metal foam. The weak mechanical properties of graphite foam block its development as a heat exchanger. Adding more material into the graphite foam or changing the fabrication process might improve the mechanical properties of the foam [31]. The major disadvantage of open-cell foams is the relatively high pressure drop penalty, as pointed out by Muley et al. [32]. Besides, at present, foam heat exchangers are not suitable for high-pressure conditions, as the maximum pressure for foam heat exchangers is relatively low.

6.7.3.2 Ceramic Materials

The two main advantages of using ceramic materials and ceramic matrix composites (CMCs) in heat exchanger construction over traditional metallic

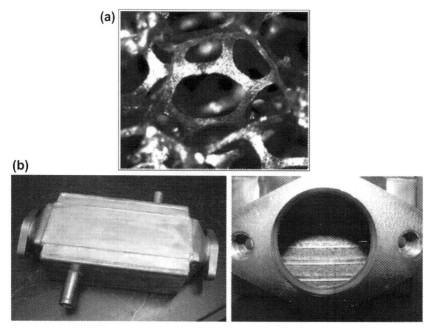

Figure 6.19 (a) Open-cell metal foam and (b) a foam heat exchanger [32].

materials are their temperature resistance and corrosion resistance. Ceramics and CMCs are now used in rocket nozzles and jet engines, heat shields for space vehicles, aircraft brakes, gas turbines for power plants, fusion reactor walls, heat treatment furnaces, heat recovery systems, etc. Major obstacles preventing the wide use of ceramics include ceramic-metallic mechanical sealing, manufacturing costs and methods, and their brittleness in tension. The most commonly used CMCs are nonoxide CMCs, i.e., carbon/carbon, carbon/silicon carbide, and silicon carbide/silicon carbide. At present, various heat exchangers, such as PFHEs, micro heat exchangers, and primary surface heat exchangers, can be manufactured by ceramics. Ceramic recuperators and intercoolers can be used to improve the gas turbine cycle efficiency. A ceramic honeycomb heat exchanger is illustrated in Fig. 6.20.

6.8 SUMMARY

Aerospace heat exchangers are mainly used in gas turbine cycles, ECSs, and thermal management of various civil and military aerospace platforms. The motivation of aerospace heat exchangers is discussed. The main features are listed for aerospace heat exchangers. This chapter also exemplifies several

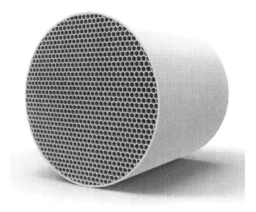

Figure 6.20 A ceramic honeycomb heat exchanger.

typical compact heat exchangers including PFHEs, micro heat exchangers, PCHEs, primary surface heat exchangers, and heat pipes, which are promising to provide efficient heat transfer and cooling in the aerospace industry.

Interested readers can find more information about aerospace heat exchangers in Refs. [33–38].

REFERENCES

[1] Sundén B. Introduction to heat transfer. Southampton: WIT Press; 2012.
[2] Shah RK, Sekulić DP. Fundamentals of heat exchanger design. John Wiley & Sons, Inc; 2003.
[3] Wang L, Sundén B, Manglik RM. Plate heat exchangers: design, applications and performance. WIT Press; 2007.
[4] Reay D, Ramshaw C, Harvey A. Process intensification: engineering for efficiency, sustainability and flexibility. Oxford: Butterworth-Heinemann; 2013.
[5] Bergles AE. ExHFT for fourth generation heat transfer technology. Exp Therm Fluid Sci 2002;26(2):335—44.
[6] Webb RL, Kim NH. Principles of enhanced heat transfer. Taylor & Francis Group; 2005.
[7] European commission, flightpath 2050 Europe's vision for aviation. 2011.
[8] Yunus AC, Michael AB. Thermodynamics: an engineering approach. New York: McGraw-Hill; 2006.
[9] Sieber J. Overview NEWAC (new aero engine core concepts). April 2009.
[10] Wright S, Andrews G, Sabir H. A review of heat exchanger fouling in the context of aircraft air-conditioning systems, and the potential for electrostatic filtering. Appl Therm Eng 2009;29(13):2596—609.
[11] North MT, Cho WL. High heat flux liquid-cooled porous metal heat sink. In: Proceedings of Interpack '03, ASME paper IPACK2003-35320, Kaanapali, HI; 2003.

[12] Marcinichen JB, Olivier JA, Thome JR. Reasons to use two-phase refrigerant cooling. Electronics Cooling Magazine; March 2011. p. 22—7.

[13] Thompson D. Cool system, hot results. Wind systems magazine; February 2010. p. 40—3.

[14] Strumpf H, Mirza Z. Development of a microchannel heat exchanger for aerospace applications. In: ASME 2012 10th International Conference on Nanochannels, Microchannels, and Minichannels collocated with the ASME 2012 Heat Transfer Summer Conference and the ASME 2012 Fluids Engineering Division Summer Meeting. American Society of Mechanical Engineers; 2012. p. 459—67.

[15] Focke WW, Zachariades J, Olivier I. The effect of the corrugation inclination angle on the thermohydraulic performance of plate heat exchangers. Int J Heat Mass Transf 1985;28(8):1469—79.

[16] Mylavarapu SK, Sun X, Christensen RN, Unocic RR, Glosup RE, Patterson MW. Fabrication and design aspects of high-temperature compact diffusion bonded heat exchangers. Nuc Eng Des 2012;249:49—56.

[17] Nicholas MG. Joining processes: introduction to brazing and diffusion bonding. Springer; 1998.

[18] Tsuzuki N, Kato Y, Ishiduka T. High performance printed circuit heat exchanger. Appl Therm Eng 2007;27(10):1702—7.

[19] Kim DE, Kim MH, Cha JE, Kim SO. Numerical investigation on thermal—hydraulic performance of new printed circuit heat exchanger model. Nuc Eng Des 2008;238(12):3269—76.

[20] Dixit T, Ghosh I. Review of micro- and Mini-channel heat sinks and heat exchangers for single phase fluids. Renew Sustain Energy Rev 2015;41:1298—311.

[21] Tuckerman DB, Pease RFW. High-performance heat sinking for VLSI. IEEE Electron Device Lett 1981;2:126—9.

[22] Wu Z, Sundén B. On further enhancement of single-phase and flow Boiling heat transfer in micro/Minichannels. Renew Sustain Energy Rev 2014;40:11—27.

[23] Sundén B, Wu Z. Advanced heat exchangers for clean and sustainable technology. In: Yan J, editor. Handbook of clean energy systems. John Wiley & Sons; 2015.

[24] Alm B, Imke U, Knitter R, Schygulla U, Zimmermann S. Testing and Simulation of ceramic micro heat exchangers. Chem Eng J 2008;135:S179—84.

[25] Morini GL. Scaling effects for liquid flows in microchannels. Heat Transf Eng 2006;27:64—73.

[26] Maranzana G, Perry I, Maillet D. Mini-and micro-channels: influence of axial conduction in the walls. Int J Heat Mass Transf 2004;47:3993—4004.

[27] Nacke R, Northcutt B, Mudawar I. Theory and experimental validation of cross-flow micro-channel heat exchanger module with reference to high Mach aircraft gas turbine engines. Int J Heat Mass Transf 2011;54(5):1224—35.

[28] Baker N. Intercooled engine and integration, European Workshop on new aero engine concepts, Munich. June 30—July 1, 2010.

[29] Wang Y, Peterson GP. Investigation of a novel flat heat pipe. ASME J Heat Transf 2005;127(2):165—70.

[30] Mahjoob S, Vafai K. A synthesis of fluid and thermal transport models for metal foam heat exchangers. Int J Heat Mass Transf 2008;51(15):3701—11.

[31] Lin WM, Sundén B, Yuan JL. A performance analysis of porous graphite foam heat exchangers in vehicles. Appl Therm Eng 2013;50(1):1201—10.

[32] Muley A, Kiser C, Sundén B, Shah RK. Foam heat exchangers: a technology assessment. Heat Transf Eng 2012;33(1):42—51.

[33] Rishmany J, Mabru C, Chieragatti R, Rezaï Aria F. Simplified modelling of the behaviour of 3D-periodic structures such as aircraft heat exchangers. Int J Mech Sci 2008;50(6):1114—22.

[34] Ito Y, Nagasaki T. Suggestion of intercooled and recuperated jet engine using already equipped components as heat exchangers. In: 47th AIAA/ASME/SAE/ASEE Joint Propusion Conference & exhibit, San Diego, California; 2011.

[35] Doo JH, Ha MY, Min JK, Stieger R, Rolt A, Son C. An investigation of cross-corrugated heat exchanger primary surfaces for advanced intercooled-cycle aero engines (Part-I: novel geometry of primary surface). Int J Heat Mass Transf 2012;55(19—20):5256—67.

[36] Shukla KN. Heat pipe for aerospace applications—an Overview. J Electron Cool Therm Control 2015;5:1—14.

[37] Oliveira JLG, Tecchio C, Paiva KV, Mantelli MBH, Gandolfi R, Ribeiro LGS. In-flight testing of loop thermosyphons for aircraft cooling. Appl Therm Eng 2016;98:144—56.

[38] Musto M, Bianco N, Rotondo G, Toscano F, Pezzella G. A Simplified methodology to simulate a heat exchanger in an Aircraft's oil cooler by means of a porous media model. Appl Therm Eng 2016;94:836—45.

CHAPTER 7

Heat Pipes for Aerospace Application

7.1 INTRODUCTION

Heat pipes are useful in the thermal control of spacecraft, satellites, and avionics. Cooling of satellite electronic components within a limited surface area represents an example. It is possible to transfer heat over long distances, with minimal temperature gradients. Commonly a heat pipe has no moving parts and high-temperature heat sinks are allowed. Aluminum heat pipes with ammonia as the working fluid are common. Heat pipes can be designed as constant conductance heat pipes, variable conductance heat pipes (VCHPs), and loop heat pipes (LHPs). The LHPs offer effective heat removal over long distances without being sensitive to gravity. LHPs may have multiple evaporators to accommodate distributed heat sources and passive or active thermal control.

An early application of heat pipes for space missions was to make the temperature of satellite transponders uniform. As satellites are on orbit, one side is exposed to direct radiation from the sun, whereas the opposite side is exposed to the deep cold outer space. This results in severe temperature gradients, affecting the reliability and accuracy of the transponders. The cooling system developed was the so-called VCHPs to actively regulate the heat flow or the evaporator temperature.

Heat pipes have attracted a huge interest among researchers and practitioners. An overview was presented by Shukla [1]. Heat pipes have become attractive components for spacecraft cooling and temperature stabilization because of their lightweight, low maintenance requirement, and reliability. As one side of the spacecraft is subject to intense solar radiation and the other is exposed to deep space, heat pipes have been used to transport heat from the side irradiated by the Sun to the cold side to maintain a more uniform temperature in the structure. Heat pipes have also been used to dissipate heat generated by electronic components in satellites [2]. Early experiments of heat pipes for aerospace applications were conducted in sounding rockets. In 1974, 10 separate heat pipe experiments

Heat Transfer in Aerospace Applications
ISBN 978-0-12-809760-1
http://dx.doi.org/10.1016/B978-0-12-809760-1.00007-7

were carried out in the International Heat Pipe Experiment [3]. Aboard the application technology satellite-6, experiments with heat pipes using an ammonia heat pipe with a spiral artery wick were carried out. The heat pipe was used as a thermal diode [4]. Using the space shuttle, flight testing of prototype heat pipe designs continued at a very large scale [5−7]. Heat pipe thermal buses were proposed to facilitate a connection between the heat-generating components and an external radiator [8−10]. In 1992, two different axially grooved oxygen heat pipes were tested aboard the shuttle discovery (STS-53) by NASA and the US Air Force to determine the startup behavior and transport capabilities during microgravity operations [11].

An advanced capillary structure, which combined reentrant and a large number of microgrooves for the heat pipe evaporator, was investigated in microgravity conditions during the so-called 2005 FOTON-M2 mission of the European Space Agency [12]. The NASA thermotechnical challenges and opportunities for space exploration, with emphasis on heat pipes and two-phase thermal loops, were presented in Ref. [13].

In a study [14], a heat pipe laser mirror was designed and fabricated to test the feasibility of this technology compared to water-cooled or un-cooled mirrors for high-power lasers. Thermal diodes have been proposed for cooling low-temperature sensors, such as an infrared detector in low subsolar Earth orbits [15]. A replacement of the radioisotope thermoelectric generating systems by radioisotope Stirling systems as a long-lasting electricity generation solution for space missions was proposed in Ref. [16]. Alkali-metal VCHPs were proposed and tested to allow multiple stops and restarts of a Stirling engine [17]. Different shapes and structures of heat pipes were studied and tested in Refs. [18−20].

The constrained vapor bubble experiment [21] provided a state-of-the-art heat pipe research undertaken by NASA to cool the International Space Station (ISS).

7.2 GENERAL DESCRIPTION OF HEAT PIPES

Passive two-phase heat transfer devices capable of transferring large quantities of heat with a minimal temperature drop were first introduced by Gaugler [22]. These devices, however, received limited attention until Grover et al. [23] published the results from an independent investigation and introduced the notation heat pipe. Since then, heat pipes have been used in a huge number of applications ranging from temperature control of the permafrost layer under the Alaska pipeline to thermal control of electronic components such as high-power semiconductor devices [24].

A conventional heat pipe was described by Rohsenow et al. [25] and it consists of a sealed container lined with a wicking structure. The container is evacuated and backfilled with an amount of liquid to fully saturate the wick. When a heat pipe operates on a closed two-phase cycle with only pure liquid and vapor present, the working fluid remains saturated as long as the operating temperature is between the freezing point and the critical state. As shown in Fig. 7.1, the heat pipe has three distinct regions, namely, the evaporator region where heat is added, the condenser region where heat is rejected, and the adiabatic or isothermal region. The added heat in the evaporator region of the container causes the working fluid in the evaporator wicking structure to vaporize. The high temperature and corresponding high pressure in this region result in flow of the vapor to the other, cooler end of the container, where the vapor condenses and releases the latent heat of vaporization. The capillary forces in the wicking structure then pump the liquid back to the evaporator.

The heat pipe container provides containment and structural stability. It must be fabricated from a material (1) that is compatible with both the working fluid and the wicking structure, (2) that is strong enough to withstand the pressure associated with the saturation temperatures encountered during storage and normal operation, and (3) that has a high thermal conductivity to permit the effective transfer of heat either into or out of the vapor space. In addition, the container material must be resistant to corrosion resulting from interaction with the environment and must be adaptive to be formed into an appropriate size and shape.

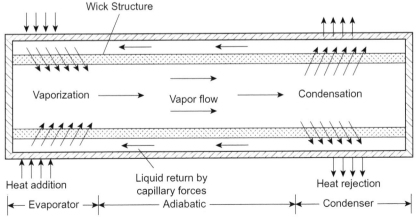

Figure 7.1 Principle of heat pipe operation.

The wicking structure has two functions during the heat pipe operation. It transports and provides the mechanism for the return of the working fluid from the condenser to the evaporator. It also ensures that the working fluid is evenly distributed over the evaporator surface. In order to provide a flow path with low flow resistance, an open porous structure with high permeability is desirable. However, to increase the capillary pumping pressure, a small pore size is necessary. Using a nonhomogeneous wick made of several different materials or a composite wicking structure may provide good function.

As the operation of a heat pipe is based on vaporization and condensation of the working fluid, selection of a suitable fluid is an important factor in the design and manufacturing of heat pipes. The most common applications involve the use of heat pipes with a working fluid having a boiling temperature between 250 and 375K. However, both cryogenic heat pipes (operating in the 5–100K temperature range) and liquid metal heat pipes (operating in the 750–5000K temperature range) have been developed and used successfully. In addition to the thermophysical properties of the working fluid, other factors such as the compatibility of the materials and the ability of the working fluid to wet the wick and wall materials must be considered [26,27]. Further criteria for the selection of the working fluids have been presented by Groll et al. [28], Peterson [29], and Faghri [30].

As heat pipes utilize a capillary wicking structure to promote the flow of liquid from the condenser to the evaporator, they can be used in a horizontal orientation, in microgravity environments, or even in applications where the capillary structure must promote the liquid against gravity from the evaporator to the condenser.

7.3 CAPILLARY LIMITATION

The heat pipe performance and operation are strongly dependent on the shape, working fluid, and wick structure, but the fundamental phenomenon that governs the operation of such devices arises from the difference in capillary pressure across the liquid–vapor interfaces in the evaporator and condenser regions. The vaporization occurring in the evaporator section causes the meniscus to recede into the wick, and condensation in the condenser section causes flooding. The combined effect of vaporization and condensation results in a meniscus radius of curvature that varies along the axial length of the heat pipe, as conjectured in Fig. 7.2a. The location where the meniscus has a minimum radius of curvature is commonly referred to as

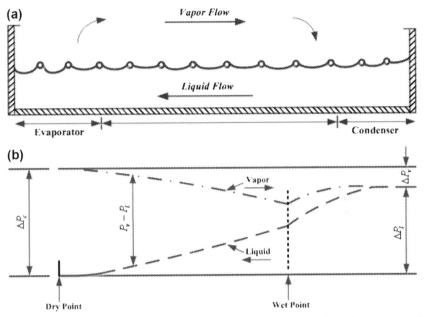

Figure 7.2 (a) Variation of meniscus curvature as a function of axial position and (b) typical liquid and vapor pressure distributions in a heat pipe. *(Based on Peterson GP. An Introduction to heat pipes: modeling, testing and applications. Washington (DC): J. Wiley & Sons, Inc.; 1994.)*

the dry point and usually occurs in the evaporator at a point farthest from the condenser region. The wet point occurs at the location where the vapor pressure and liquid pressure are approximately equal or where the radius of curvature is maximum. However, this location can be anywhere in the condenser or adiabatic section but is typically found near the end of the condenser farthest away from the evaporator, as reported in Ref. [29].

Fig. 7.2b illustrates the relationship between the static liquid and static vapor pressures in an operating heat pipe. The capillary pressure gradient across a liquid–vapor interface is equal to the pressure difference between the liquid and vapor phases at any given axial position. For a heat pipe to function properly, the net capillary pressure difference between the wet and dry points (Fig. 7.2b) must be greater than the sum of all pressure losses occurring throughout the liquid and vapor flow paths. This relationship, referred to as the capillary limitation, can be expressed mathematically as

$$(\Delta P_c)_m \geq \int_{L_{eff}} \frac{\partial P_v}{\partial x} dx + \int_{L_{eff}} \frac{\partial P_l}{\partial x} dx + \Delta P_{PT,e} + \Delta P_{PT,c} + \Delta P_+ + \Delta P_\| \quad (7.1)$$

where

$(\Delta P_c)_m$ = the maximum capillary pressure difference generated within the capillary wicking structure between the wet and dry points;

$\partial P_v/\partial x$ = sum of the inertial and viscous pressure drop occurring in the vapor phase;

$\partial P_l/\partial x$ = sum of the inertial and viscous pressure drop occurring in the liquid phase;

$\Delta P_{PT,e}$ = pressure gradient across the phase transition in the evaporator;

$\Delta P_{PT,c}$ = pressure gradient across the phase transition in the condenser;

ΔP_+ = normal hydrostatic pressure drop;

$\Delta P_{\parallel}$ = axial hydrostatic pressure drop.

The first two terms on the right hand side of Eq. (7.1), $\partial P_v/\partial x$ and $\partial P_l/\partial x$, represent the sum of viscous and inertial losses in the vapor and liquid flow paths, respectively. The next two terms, $\Delta P_{PT,e}$ and $\Delta P_{PT,c}$, represent the pressure gradients occurring across the phase transition in the evaporator and condenser, but can often be neglected. The last two terms, ΔP_+ and $\Delta P_{\parallel}$, represent the normal and axial hydrostatic pressure drops, respectively. When the maximum capillary pressure is equal to or greater than the sum of the normal and axial hydrostatic pressure drops, the capillary structure is capable of returning an adequate amount of the working fluid to prevent dry out of the evaporator wicking structure. If the total capillary pressure across the liquid−vapor interface is not greater than or equal to the sum of all the pressure drops occurring throughout the liquid vapor flow paths, the working fluid will not be returned to the evaporator, causing the liquid in the evaporator wicking structure to vanish, leading to dry out. This condition, referred to as the capillary limitation, depends on the wicking structure, working fluid, evaporator heat flux, and operating temperature. The following sections give a brief description of the individual pressure loss terms based on the information given by Bar-Cohen and Kraus [24] and Peterson and Feng [27].

7.3.1 Capillary Pressure

At the surface of a single liquid−vapor interface, a capillary pressure difference, defined as $(P_v - P_l)$ or ΔP_c, exists. This capillary pressure difference is described mathematically by the Laplace−Young equation, i.e.,

$$\Delta P_c = \sigma \left(\frac{1}{r_1} + \frac{1}{r_2} \right) \tag{7.2}$$

where r_1 and r_2 are the principal radii of curvature and σ is the surface tension. For many heat pipe wicking structures the maximum capillary pressure may be written in terms of a single radius of curvature, r_c. Then the maximum capillary pressure between the wet and dry locations can be expressed as the difference between the capillary pressure across the meniscus at the wet location and the capillary pressure at the dry location, or as

$$\Delta P_{c,m} = \left(\frac{2\sigma}{r_{c,e}}\right) - \left(\frac{2\sigma}{r_{c,c}}\right) \tag{7.3}$$

Fig. 7.2a illustrates the effect of the vaporization occurring in the evaporator, which causes the liquid meniscus to recede into the wick, and the condensation occurring in the condenser section, which causes flooding of the wick. This combination of meniscus recession and flooding results in a reduction in the local capillary radius $r_{c,e}$ and increases the local capillary radius $r_{c,c}$, respectively, which further results in a pressure difference and accordingly pumping of the liquid from the condenser to the evaporator. During steady-state operation, it is generally assumed that the capillary radius in the condenser or at the wet location $r_{c,c}$ approaches infinity, so that the maximum capillary pressure for a heat pipe operating at steady state can be expressed as a function of only the effective capillary radius of the evaporator wick [29]

$$\Delta P_{c,m} = \left(\frac{2\sigma}{r_{c,e}}\right) \tag{7.4}$$

Values for the effective capillary radius r_c are provided in Table 7.1 [29] for some common wicking structures. In the case of other geometries, the effective capillary radius can be found theoretically using the methods proposed by Chi [31] or experimentally using the methods described by Ferrell and Alleavitch [32], Freggens [33], or Tien [34]. In addition, limited information on the transient behavior of capillary structures is available in Ref. [35].

7.3.2 Normal Hydrostatic Pressure Drop

Two hydrostatic pressure drop parts are of importance in heat pipes, namely, a normal hydrostatic pressure drop ΔP_+, which occurs only in heat pipes that have circumferential communication of the liquid in the wick, and an axial hydrostatic pressure drop. The first one is the result of the body force component acting perpendicularly to the longitudinal axis of the heat pipe and can be expressed as

$$\Delta P_+ = \rho_l g d_v \cos\psi \tag{7.5}$$

Table 7.1 Effective Capillary Radius of Several Wick Structures

Structure	r_c	Data
Circular cylinder (artery or tunnel wick)	r	Radius of liquid flow passage
Rectangular groove	ω	Groove width
Triangular groove	$\omega/\cos\beta$	β = half include angle
Parallel wires	ω	Wire spacing
Wire screens	$(\omega + d_w)/2 = 1/2N$	d = wire diameter N = screen mesh number
Packed spheres	$0.41r_s$	r_s = sphere radius

Based on Peterson GP. An Introduction to heat pipes: modeling, testing and applications. Washington (DC): J. Wiley & Sons, Inc.; 1994.

where ρ_l is the density of the liquid; g, the gravitational acceleration; d_v, the diameter of the vapor portion of the pipe; and ψ, the angle the heat pipe makes with respect to the horizontal.

7.3.3 Axial Hydrostatic Pressure Drop

The axial hydrostatic pressure drop, $\Delta P_\parallel$, results from the component of the body force acting along the longitudinal axis. This term can be expressed as

$$\Delta P_\parallel = \rho_l gL \sin\psi \qquad (7.6)$$

where L is the overall length of the heat pipe.

In an environment with gravity, the normal and axial hydrostatic pressure drop parts may either assist or hinder the capillary pumping process depending on whether the inclination of the heat pipe promotes or hinders the flow of liquid back to the evaporator (i.e., the evaporator is placed either below or above the condenser). In a zero-gravity environment, both the parts can be neglected because of the absence of body forces.

7.3.4 Liquid Pressure Drop

The capillary pumping pressure promotes the flow of liquid through the wicking structure but the viscous forces in the liquid result in a pressure drop ΔP_l, which resists the capillary flow through the wick. This liquid pressure gradient may vary along the longitudinal axis of the heat pipe, and hence the total liquid pressure drop might be determined by integrating the pressure gradient over the length of the flow passage [28], or

$$\Delta P_l(x) = -\int_0^x \frac{dP_l}{dx}dx \qquad (7.7)$$

where the limits of integration are from the evaporator end to the condenser end ($x = 0$) and dP_l/dx is the gradient of the liquid pressure resulting from frictional drag. Introduction of the Reynolds number Re_l and drag coefficient f_l and substituting the local liquid velocity, which is related to the local heat flow, the wick cross-sectional area, the wick porosity ε, and the latent heat of vaporization λ, yield

$$\Delta P_l = \left(\frac{\mu_l}{KA_w\lambda\rho_l} \right) L_{eff} q \qquad (7.8)$$

where L_{eff} is the effective heat pipe length, which is defined as

$$L_{eff} = 0.5L_e + L_a + 0.5L_c \qquad (7.9)$$

and the wick permeability K is given in Table 7.2.

7.3.5 Vapor Pressure Drop

Addition of mass and mass removal in the evaporator and condenser, respectively, along with the compressibility of the vapor phase, complicate the vapor pressure drop in heat pipes. Applying the continuity condition to the adiabatic region of the heat pipe ensures that for continued operation, the liquid mass flow rate and vapor mass flow rate must be equal. Because of the difference in density between these two phases, the vapor

Table 7.2 Wick Permeability of Several Structures

Structure	K	Data
Circular cylinder (artery or tunnel wick)	$r^2/8$	r = radius of liquid flow passage
Open rectangular grooves	$2\varepsilon(r_{h,l})^2/(f_l Re_l)$	ε = wick porosity w = groove width s = groove pitch δ = groove depth $(r_{h,l}) = 2\omega\delta/(\omega + 2\delta)$
Circular annular wick	$2(r_{h,l})^2/(f_l Re_l)$	$(r_{h,l}) = r_1 - r_2$
Wrapped screen wick	$\dfrac{d_w^2\varepsilon^3}{122(1-\varepsilon)^2}$	d_w = wire diameter $\varepsilon = 1 - (1.05\pi N d_w/4)$ N = mesh number
Packed sphere	$r_s^2\varepsilon^3/37.5(1-\varepsilon)^2$	r_s = sphere radius ε = porosity (dependent on packing mode)

From Peterson GP. An Introduction to heat pipes: modeling, testing and applications. Washington (DC): J. Wiley & Sons, Inc.; 1994.

velocity is significantly higher than the velocity of the liquid phase. For this reason, in addition to the pressure gradient resulting from frictional drag, the pressure gradient due to variations in the dynamic pressure must also be considered. Chi [31], Dunn and Reay [35], and Peterson [29] all addressed this problem. The results indicate that on integration of the vapor pressure gradient, the dynamic pressure effects cancel. The result is an expression that is similar to that developed for the liquid,

$$\Delta P_v = \left(\frac{C(f_v Re_v)\mu_v}{2(r_{h,v})^2 KA_v \rho_v \lambda} \right) L_{eff} q \tag{7.10}$$

where $(r_{h,v})$ is the hydraulic radius of the vapor space and C is a constant that depends on the Mach number.

During steady-state operation, the liquid mass flow rate m must be equal to the vapor mass flow rate m, at every axial position, and although the liquid flow regime is always laminar, the vapor flow might be either laminar or turbulent. As a result, the vapor flow regime must be written as a function of the heat flux. It is also necessary to determine whether the flow is compressible or incompressible by evaluating the local Mach number.

A number of early investigations summarized by Bar-Cohen and Kraus [24] have shown that the following conditions can be used with reasonable accuracy.

$$Re_v < 2300, \quad Ma_v < 0.2$$
$$(f_v Re_v) = 16 \tag{7.11}$$
$$C = 1.00$$

$$Re_v < 2300, \quad Ma_v > 0.2$$
$$(f_v Re_v) = 16$$
$$C = \left[1 + \left(\frac{\gamma_v - 1}{2} \right) Ma_v^2 \right]^{-\frac{1}{2}} \tag{7.12}$$

$$Re_v > 2300, \quad Ma_v < 0.2$$
$$(f_v Re_v) = 0.038 \left(\frac{2(r_{h,v})q}{A_v \mu_v \lambda} \right)^{\frac{3}{4}} \tag{7.13}$$
$$C = 1.00$$

$$\text{Re}_v > 2300, \quad \text{Ma}_v > 0.2$$

$$(f_v \text{Re}_v) = 0.038 \left(\frac{2(r_{h,v})q}{A_v \mu_v \lambda} \right)^{\frac{3}{4}} \tag{7.14}$$

$$C = \left[1 + \left(\frac{\gamma_v - 1}{2} \right) \text{Ma}_v^2 \right]^{-\frac{1}{2}}$$

Because the equations used to evaluate both the Reynolds number and the Mach number are functions of the heat transport capacity, the conditions of the vapor flow must first be assumed. Using these assumptions, the maximum heat capacity $q_{c,m}$ can be determined by substituting the values of the individual pressure drops in Eq. (7.1) and solving for $q_{c,m}$. Once the value of $q_{c,m}$ is known, it can then be substituted into the expressions for the vapor Reynolds number and the Mach number to determine the accuracy of the original assumption. Using this iterative approach [31], accurate values for the capillary limitation as a function of the operating temperature can be determined for $(qL)_{c,m}$ (in W m) and $q_{c,m}$ (in W).

7.4 OTHER LIMITATIONS

In addition to the capillary limitation, there are several other important mechanisms that limit the maximum amount of heat transferred during steady-state operation of a heat pipe. Among these mechanisms are the viscous limit, sonic limit, entrainment limit, and boiling limit. The capillary wicking limit and viscous limits are related to the pressure drops occurring in the liquid and vapor phases, respectively. The sonic limit is established by the choked flow occurring in the vapor passage, whereas the entrainment limit is a result of the high liquid—vapor shear forces developed when the vapor passes in counterflow over the liquid saturated wick. The boiling limit is reached when the heat flux applied in the evaporator portion is so high that nucleate boiling occurs in the evaporator wick. This creates vapor bubbles that partially block the return of the liquid.

In low-temperature applications using cryogenic working fluids, either the viscous limit or capillary limit occurs first. In high-temperature heat pipes using, e.g., liquid metal working fluids, the sonic and entrainment limits are of increased significance. The theory and fundamental phenomena that cause these limitations have been the objectives of quite many investigations and these are well documented by Chi [31], Dunn and Reay

[35], Tien [34], Peterson [29], Faghri [30], and proceedings from the International Heat Pipe Conference series.

7.4.1 Viscous Limitation

When the operating temperature is very low, the vapor pressure difference between the closed end of the evaporator (the high-pressure region) and the closed end of the condenser (the low-pressure region) might be very small. Accordingly the viscous forces within the vapor region might dominate and hence limit the heat pipe operation. Dunn and Reay [35] discussed this limit in more detail and suggested the criterion

$$\frac{\Delta P_v}{P_v} < 0.1 \tag{7.15}$$

for determining when this limit is of major concern.

7.4.2 Sonic Limitation

The sonic limit in heat pipes is similar to the sonic limit that occurs in converging-diverging nozzles [29], except that in a converging-diverging nozzle the mass flow rate is constant and the vapor velocity varies because of the change in cross-sectional area, whereas in heat pipes the area is constant and the vapor velocity varies because of the evaporation and condensation along the heat pipe. Similar to nozzle flow, a decreased outlet pressure, or, in this case, condenser temperature, results in a decrease in the evaporator temperature until the sonic limit is reached. A further increase in the heat rejection rate does not reduce the evaporator temperature or the maximum heat transfer capacity, but reduces the condenser temperature because of the choked flow.

The sonic limitation in heat pipes can be determined by

$$q_{s,m} = A_v \rho_v \lambda \left(\frac{\gamma_v R_v T_v}{2(\gamma_v + 1)} \right)^{\frac{1}{2}}, \tag{7.16}$$

where T_v is the mean vapor temperature within the heat pipe.

7.4.3 Entrainment Limitation

Because of the high vapor velocities, liquid droplets can be picked up or entrained in the vapor flow and this results in excess liquid accumulation in the condenser and dry out of the evaporator wick [36]. For proper operation the onset of entrainment in countercurrent two-phase flow must be avoided. The most commonly quoted criterion to determine this onset is based on the Weber number, We, defined as the ratio of the viscous shear force to the force resulting from the liquid surface tension, i.e.,

$$We = \frac{2(r_{h,w})\rho_v V_v^2}{\sigma} \qquad (7.17)$$

If the Weber number is equal to unity, the onset of entrainment of liquid droplets in the vapor flow is reached. Thus the Weber number must be less than unity.

By relating the vapor velocity to the heat transport capacity, a value for the maximum heat transport capacity based on the entrainment limitation can be determined as

$$V_v = \frac{q}{A_v \rho_v \lambda} \qquad (7.18)$$

$$q_{e,m} = A_v \lambda \left(\frac{\sigma \rho_v}{2(r_{h,w})}\right)^{-\frac{1}{2}} \qquad (7.19)$$

where $(r_{h,w})$ is the hydraulic radius of the wick structure, defined as twice the area of the wick pore at the wick—vapor interface divided by the wet perimeter at the wick—vapor interface. Rice and Fulford [37] developed a different approach resulting in an expression to define the critical dimensions for wicking structures to prevent entrainment.

7.4.4 Boiling Limitation

At very high radial heat fluxes, nucleate boiling may occur in the wicking structure and bubbles may become trapped in the wick, blocking the liquid return and resulting in dry out in the evaporator. This phenomenon, referred to as the boiling limit, differs from the other limitations previously presented because it depends on the evaporator heat flux as opposed to the axial heat flux [29].

The boiling limit is found by examining the nucleate boiling theory. This involves two separate phenomena: bubble formation and the subsequent growth or collapse of the bubbles. The first phenomenon, bubble formation, is governed by the number and size of nucleation sites on a solid surface. The second one, bubble growth or collapse, depends on the liquid temperature and corresponding pressure caused by the vapor pressure and surface tension of the liquid. Utilizing a pressure balance on a bubble and applying the Clausius—Clapeyron equation to relate the temperature and pressure, an expression for heat flux, beyond which bubble growth will occur, can be developed, i.e., [31]

$$q_{b,m} = \left(\frac{2\pi L_{eff} k_{eff} T_v}{\lambda \rho_v \ln(r_i/r_v)}\right)\left(\frac{2\sigma}{r_n} - \Delta P_{c,m}\right) \qquad (7.20)$$

where k_{eff} is the effective thermal conductivity of the liquid—wick combination. Such values are provided in Table 7.3. r_i is the inner radius of the heat pipe wall and r_n is the nucleation site radius, which according to Dunn and Reay [35], can be assumed to range from 2.54×10^{-5} to 2.54×10^{-7} m for conventional heat pipes.

As the power level associated with each of the four mentioned limitations has been determined as a function of the maximum heat transport capacity, the operating envelope can be determined. Then it is a matter of selecting the lowest limitation for any given operating temperature to determine the heat transport limitation applicable for a prespecified set of conditions.

7.5 DESIGN AND MANUFACTURING CONSIDERATIONS FOR HEAT PIPES

Several studies have focused on problems associated with the design and manufacturing of heat pipes. Among these are the works by Feldman [38], Brennan and Kroliczek [39], Peterson [29], and Faghri [30]. Besides factors such as cost, size, weight, reliability, working fluid, and construction and sealing techniques, the design and manufacturing of heat pipes are governed by three operational factors, namely, the effective operating temperature range (determined by the selection of the working fluid), the maximum power the heat pipe is capable of transporting (determined by the ultimate pumping capacity of the wick structure), and the maximum

Table 7.3 Effective Thermal Conductivity of Liquid-Saturated Wick Structures

Wick Structure	k_{eff}
Wick and liquid in series	$\dfrac{k_l k_w}{\varepsilon k_w + k_l(1-\varepsilon)}$
Wick and liquid in parallel	$\varepsilon k_l + k_w(1-\varepsilon)$
Wrapped screen	$\dfrac{k_l[(k_l+k_w) - (1-\varepsilon)(k_l - k_w)]}{[(k_l+k_w) + (1-\varepsilon)(k_l - k_w)]}$
Packed spheres	$\dfrac{k_l[(2k_l+k_w) - 2(1-\varepsilon)(k_l - k_w)]}{[(2k_l+k_w) + (1-\varepsilon)(k_l - k_w)]}$
Rectangular grooves	$\dfrac{(\omega_f k_l k_w \delta) + \omega k_l(0.815\omega_f k_w + \delta k_l)}{(\omega + \omega_f)(0.815\omega_f k_l + \delta k_l)}$

From Peterson GP. An Introduction to heat pipes: modeling, testing and applications. Washington (DC): J. Wiley & Sons, Inc.; 1994; Chi SW, Heat pipe theory and practice, New York: McGraw-Hill Publishing Company; 1976.

evaporator heat flux (determined by the position where nucleate boiling occurs).

7.5.1 Selection of Working Fluid

Because heat pipes rely on vaporization and condensation to transfer heat, selection of a suitable working fluid is an important factor. For most moderate temperature applications, working fluids with boiling temperatures between 250 and 375K are required. Possible fluids are ammonia, Freon-11 (trichlorofluoromethane) or Freon-113 (trichlorotrifluoroethane), acetone, methanol, and water. For a capillary-wick-limited heat pipe, the characteristics of a good working fluid are high latent heat of vaporization, high surface tension, high liquid density, and low liquid viscosity. Chi [31] combined these properties into a parameter, N_l, referred to as the liquid transport factor or figure of merit, defined as

$$N_l = \frac{\rho_l \sigma \lambda}{\mu_l} \tag{7.21}$$

This parameter can be used to evaluate various working fluids at specific operating temperatures. The concept of a single parameter for evaluating working fluids was extended by Gosse [40], who demonstrated that the thermophysical properties of the liquid—vapor equilibrium state could be reduced to three independent parameters.

Along with the importance of the thermophysical properties of the working fluid, consideration must be given to the ability of the working fluid to wet the wick and wall materials, as discussed by Peterson [29]. Other important criteria in the selection of the working fluid were presented by Heine and Groll [41].

7.5.2 Importance of the Wicking Structures

In addition to providing the pumping of the liquid from the condenser to the evaporator, the wicking structure ensures that the working fluid is evenly distributed over the evaporator surface. To provide a flow path with low flow resistance through which the liquid can be returned from the condenser to the evaporator, an open porous structure with high permeability is required. However, to increase the capillary pumping pressure, a small pore size is necessary. To manage these contradictory effects, a nonhomogeneous wick made of several different materials or a composite wicking structure might be used. Udell and Jennings [42] proposed and formulated a model for a heat pipe with a wick consisting of a porous media of two different permeabilities oriented parallel to the direction of the heat flux. This wick structure provided a large pore size in the center of the wick for liquid flow and a smaller pore size for capillary pressure.

Composite wicking structures accomplish a similar effect, as the capillary pumping and axial fluid transport are handled independently. In addition to fulfilling this dual purpose, several wick structures physically separate the liquid and vapor flow. This results from an attempt to eliminate the viscous shear force that occurs during countercurrent liquid and vapor flows.

7.5.3 Compatibility of Materials

Formation of noncondensable gases (NCGs) through chemical reactions between the working fluid and the wall or wicking structure, or decomposition of the working fluid, may cause problems in the operation of the heat pipe. For these reasons, careful consideration must be given to the selection of working fluids and wicking and wall materials to prevent the occurrence of such problems during the operational life time of the heat pipe. The formation of NCG may result in either decreased performance or total failure. Corrosion problems can lead to physical degradation of the wicking structure because solid particles carried to the evaporator wick and deposited there will most likely reduce the wick permeability [43].

Basiulis et al. [44] conducted extensive compatibility tests with several combinations of working fluids and wicking structures. The findings are summarized in Table 7.4 together with investigations by Busse et al. [45]. Other investigations have also been performed by, e.g., Zaho et al. [46], in which the compatibility of water and mild steel heat pipes was evaluated; Roesler et al. [47], who evaluated stainless steel, aluminum, and ammonia combinations; and Murakami and Arai [48], who developed a statistical predictive technique for evaluating the long-term reliability of copper—water heat pipes. These studies provided additional insight into the compatibility of various liquid—material combinations. Most of the data available are based on accelerated lifetime tests.

The two problems, i.e., NCG generation and corrosion, are only two of the factors to be considered when selecting heat pipe wicks and working fluids. Other problems include wettability of the fluid—wick combination, strength-to-weight ratio, thermal conductivity and stability, and fabrication costs.

7.5.4 Sizes and Shapes of Heat Pipes

Heat pipes vary in both size and shape, ranging from a 15-m-long monogroove heat pipe developed by Alario et al. [49] for spacecraft heat

Table 7.4 Working Fluid, Wick, and Container Compatibility Data

Material	Water	Acetone	Ammonia	Methanol
Copper	RU	RU	NU	RU
Aluminum	GNC	RL	RU	NR
Stainless steel	GNT	PC	RU	GNT
Nickel	PC	PC	RU	RL
Refrasil	RU	RU	RU	RU
Material	Dow-A	Dow-E	Freon-11	Freon-113
Copper	RU	RU	RU	RU
Aluminum	UK	NR	RU	RU
Stainless steel	RU	RU	RU	RU
Nickel	RU	RL	UK	UK
Refrasil	RU		UK	UK

GNC, generation of gas at all temperatures; GNT, generation of gas at elevated temperatures when oxide is present; NR, not recommended; NU, not used; PC, probably compatible; RL, recommended by literature; RU, recommended by past successful usage; UK, unknown.
From Basiulis A, Prager RC, Lamp. Compatibility and reliability of heat pipe materials, Proc. 2nd Int. Heat pipe Conf., Bologna, Italy; 1976. pp. 357–372; Busse CA, Campanile A, Loens J. Hydrogen generation in water heat pipes at 250°C, First Int. Heat pipe Conf., Stuttgart, Germany, paper no. 4–2, October 1973.

rejection to a 10-mm-long expandable bellows type heat pipe developed by Peterson [50] for the thermal control of semiconductor devices. The cross-sectional areas of the vapor and liquid flows also vary significantly from those encountered in flat-plate heat pipes, which have very large flow areas, to commercially available heat pipes with a cross-sectional area of less than 0.30 mm^2. Heat pipes may be fixed or variable in length and either rigid or flexible for situations in which relative motion or vibration is concerned.

7.5.5 Reliability and Lifetime Tests

Peterson [29] presented an extensive review of life testing and reliability. The review indicated that many of the early investigations, e.g., those conducted by Basiulis and Filler [51] and Busse et al. [45], focused on the reliability of various types of material combinations in the intermediate operating temperature range. Long-duration life tests on copper–water heat pipes have been performed by numerous investigators. Among them, Kreed et al. [52] indicated that this combination, with proper cleaning and charging procedures, can produce heat pipes with an expected lifetime of decades. Other tests reported in Ref. [35] have indicated similar results for copper–acetone and copper–methanol combinations. However, these tests also stated that care must be taken to ensure the purity of the working fluid, wick structure, and case materials. Table 7.4 presented a summary of compatibility data obtained from various investigations.

In addition to the use of compatible materials, long-term reliability can be ensured by careful inspection and preparation processes such as

- laboratory inspection to ensure that material of high purity is used for the case, end caps, and fill tubes,
- appropriate inspection procedures to ensure that the wicking material is made from high-quality substances,
- inspection and distillation procedures to ensure that the working fluid is of high purity,
- fabrication in a clean environment to ensure or eliminate the presence of oils, vapors, etc.,
- use of clean solvents during the rinse process [53].

The effects of long-term exposure to elevated temperatures and repeated thermal cycling on heat pipes can be estimated by using a model developed by Baker [54]. This model utilizes the Arrhenius expression to predict the response parameter F,

$$F = C \, e^{-A/kT} \qquad (7.22)$$

where C is a constant; A, the reaction activation energy; k, the Boltzmann constant; and T, the absolute temperature.

This model utilizes experimental results obtained from the Jet Propulsion Laboratory in the United States to predict the rate and amount of hydrogen gas generated over a 20-year lifetime for a stainless steel heat pipe, with water as the working fluid. With this model, the generated mass of hydrogen can be predicted as a function of time.

Experiences with a wide variety of applications ranging from consumer electronics to industrial equipment have demonstrated that mechanical cleaning of the case and wicking structure with an appropriate solvent, combined with an acidic etch and vacuum bake out at elevated temperatures, can make heat pipes free of contaminants, thus enabling negligible performance degradation (less than 5%) over a product lifetime of 10 years [53].

7.6 VARIOUS TYPES OF HEAT PIPES

A heat pipe must possess vapor—liquid equilibrium with the saturated liquid and its vapor (gas phase). The saturated liquid vaporizes and moves to the condenser, where it is cooled and transferred to saturated liquid again. In a conventional heat pipe, the condensate is returned to the evaporator using a wick structure exerting a capillary action on the liquid phase of the working fluid. Various types of wick structures are used in heat pipes, including sintered metal powder, screen, and grooved wicks, which have a series of grooves parallel to the pipe axis. The performance of the heat pipes also depends on the selection of a container, a wick, and welding materials compatible with one another and with the working fluid of interest. Performance can be degraded and failures can occur in the container wall if the constituents are not compatible. For example, the constituents can react chemically or set up a galvanic cell within the heat pipe. Additionally, the container material may be soluble in the working fluid or may catalyze the decomposition of the working fluid on reaching a particular temperature limit of the working fluid. Faghri [55] provided up-to-date information on the compatibility of metals with the working fluids, which is summarized in Table 7.5. High-quality arterial grooved heat pipes are preferred for thermal stabilization of satellites. Various configurations of heat pipes are available in the market for a variety of applications [56]. Most heat pipes are generally circular cylinders. Other shapes such as rectangular (vapor chamber), conical [rotating heat pipes (RHPs)], triangular (micro heat pipes), and nose cap geometries (leading edge cooling) have also been studied.

7.6.1 Heat Pipes with Variable Conductance

A VCHP is a capillary-driven heat pipe in which an NCG is supplied to the heat pipe, in addition to the working fluid. When the VCHP is operating, the NCG is swept toward the condenser end of the heat pipe by the flow of the working fluid vapor condensing in the condenser. The NCG then blocks the working fluid from reaching a portion of the condenser. The VCHP works by variation of the portion of the condenser being available for the working fluid. As the evaporator temperature increases, the vapor temperature (and pressure) rises, the NCG is compressed, and the condenser is more exposed to the working fluid. This increases the conductivity of the heat pipe and decreases the temperature of the evaporator. Conversely, if the evaporator is cooled, the vapor pressure drops and the NCG expands. This reduces the portion of the condenser available for condensation and thus the heat pipe conductivity is decreased, which helps maintain the evaporator temperature. The first application of the VCHP to communication technology satellite was reported in [57].

Many models have been developed based on a flat front model [58], steady diffusive interface models [59—61], and a transient diffusive interface model [62] of the interface between the vapor and the NCG. The diffusive

Table 7.5 Materials Relative to Working Fluid

Working Fluid	Compatible Material	Incompatible Material
Water	Stainless steel, copper, silica, nickel, titanium	Aluminum, Inconel
Ammonia	Aluminum, stainless steel, iron, nickel, cold rolled steel	
Methanol	Stainless steel, iron, copper, brass, silica, nickel	Aluminum
Acetone	Aluminum, stainless steel, copper, brass, silica	
Freon-11	Aluminum	
Freon-21 (dichlorofluoromethane)	Aluminum, iron	
Freon-113	Aluminum	
Heptane	Aluminum	
Dowtherm	Stainless steel, copper, silica	
Lithium	Tungsten, Tantalum, molybdenum, niobium	Stainless steel, nickel, Inconel, titanium
Sodium	Stainless steel, nickel, Inconel, niobium	Titanium
Cesium	Titanium, niobium, stainless steel	
Mercury	Stainless steel	Molybdenum, Inconel, nickel, tantalum, titanium, niobium
Lead	Tungsten, tantalum	Stainless steel, nickel, Inconel, titanium, niobium
Silver	Tungsten, tantalum	Rhenium

From Faghri A. Heat pipes, review, opportunities and challenges. Front Heat Pipes (FHP), 5, 1–48, 2014.http://dx.doi.org/10.5098/fhp.5.1.

interface model assumes transient one-dimensional mass diffusion across the vapor—gas interface, with constant properties. Although the model is simplified by ignoring the compressibility in the vapor flow, it may well predict the transient operation of the VCHP. Two studies [63,64] presented a CFD model for unsteady two-dimensional heat and mass transfer in the vapor—gas region of a gas-loaded heat pipe to predict the behavior of the startup transient in the vapor—gas region. Two-dimensional transient operation of a VCHP was studied in [65].

7.6.2 Rotating Heat Pipes

In [66], a two-phase heat transfer device designed to cool machinery by removing heat through a rotating shaft was reported. The device was called an RHP. The heat input to the evaporator vaporizes the working fluid. As in ordinary heat pipes, the vapor travels down the heat pipe to the condenser, where heat is removed as the vapor condenses. In contrast to a normal heat pipe using a wick to return the condensate, an RHP uses the centrifugal force. A copper–water RHP with a copper screen mesh wick at various heat loads was tested in [67]. An experimental test rig with a water-cooled condenser section was built to study the heat transfer in the RHP for various heat loads and various rotational speeds ranging from 1000 to 2000 rpm.

7.6.3 Cryogenic Heat Pipes

The continuous growth of space-based communications and sensors, along with the evolution of aerospace and avionics, increases the demands for thermal control and heat removal in low-temperature environments. An early review of the application of cryogenic heat pipes in spacecraft was presented in [68]. Charlton and Bowman [69] developed a mathematic model to predict the performance of cryogenic heat pipes under transverse vibration. Supercritical startup behavior of cryogenic heat pipes was studied in [70]. Bughy et al. [71] discussed the thermal switching cryogenic heat pipes for thermal management of CCD cameras used in the NASA space interferometry mission.

7.6.4 Vapor Chamber

The vapor chamber is a capillary-driven planar (flat-plate heat pipe) design with a small aspect ratio. The main advantage of the vapor chamber is that it can be placed directly beneath the heat-generating avionics components without adding additional thermal resistance. There are two main applications for vapor chambers [72–74]. One application is that they are used when high powers and heat fluxes are applied to a relatively small evaporator. Heat input to the evaporator vaporizes the liquid, which flows in two dimensions to the condenser surface. After the vapor has condensed on the condenser surface, capillary forces in the wick return the condensate to the evaporator. Note that most vapor chambers are insensitive to gravity and will still operate when inverted, with the evaporator above the condenser. Another application is that compared to a one-dimensional tubular heat pipe, the width of a two-dimensional heat pipe allows an adequate cross section for heat flow, even with a very thin device.

7.6.5 Loop Heat Pipes

LHPs are two-phase heat transfer devices using capillary action to remove heat from a source and to passively move it to a condenser or radiator [75]. LHPs are similar to heat pipes but have the advantages of providing reliable operation over long distances and having the ability to operate against gravity. A large heat load can be transported over a long distance with a small temperature difference. The main components of LHPs are an evaporator, a condenser, a vapor line, a liquid line and a hydroaccumulator. A wick is required only in the evaporator and hydroaccumulator. The rest of the loop is made of smooth walled tubing. The hydroaccumulator is usually called a compensation chamber. The principle of LHPs is that the liquid in the evaporator evaporates by the applied heat load and a meniscus is formed in the liquid—vapor interface in the wick. A pressure gradient is developed by the surface tension that moves the vapor toward the condenser where it condenses. The liquid is pushed back to the evaporator by the same surface tension. The compensation chamber is connected to the evaporator by a secondary wick. With a slight variation in the place-ment of the compensation chamber wherein it is located remotely from the evaporator, a different version of the LHP is named as a capillary pumped loop (CPL). In a CPL, the compensation chamber is known as a reservoir and is outside the path of the fluid circulation. Ku [76] presented operating characteristics of an LHP. The heat transfer mechanism in the evaporator of an LHP was investigated in [77] and [78]. Shukla [79] studied the ther-mofluid dynamics of LHP operation. Different designs of LHPs ranging from powerful, large-sized LHPs to miniature LHPs (micro LHPs) have been developed and employed in a variety of applications, both in ground-based and space applications [80—82]. LHPs operating with ammonia as the working fluid are popular thermal control devices for high-powered telecommunication satellites.

7.6.6 Micro Heat Pipes

Micro heat pipes for cooling electronic devices were proposed in [83]. A micro heat pipe is defined as a heat pipe in which the mean curvature of the liquid—vapor interface is comparable in magnitude to the reciprocal of the hydraulic radius of the total flow channel. Typically, micro heat pipes have convex but cusped cross sections (e.g., a polygon), with a hydraulic diameter in the range of $10-500$ μm. A miniature heat pipe is defined as a heat pipe with a hydraulic diameter in the range of $0.5-5$ mm. An overview of the development of micro heat pipes was presented in [84]. The fabrication and experimental data on the performance characteristics of the flat water—copper heat pipe with external dimensions $2 \times 7 \times 120$ mm have been reported with radial heat fluxes of 90 and 150 W/cm^2 for horizontal and vertical applications, respectively [85]. Shukla [86] investigated the heat transfer limitations of micro heat pipes and found that the maximum heat transfer capacity of a micro heat pipe depends on the capillary and fluid continuum limits. It was found that methanol is a better suited working fluid for a micro heat pipe with triangular cross section. The capillary limit calculated in [86] was almost twice the value obtained in [87].

7.6.7 Nanofluids in Heat Pipe Applications

The working fluids in heat pipes can be helium and nitrogen for cryogenic temperatures $(2-4$ K) and liquid metals such as mercury $(523-923$ K), sodium $(873-1473$ K), and indium $(2000-3000$ K) for extremely high temperatures. The working fluid in a heat pipe is chosen according to the temperatures at which the heat pipe should operate. The vast majority of heat pipes for spacecraft and electronics cooling use ammonia $(213-373$ K), alcohols [such as methanol $(283-403$ K) or ethanol $(273-403$ K)], or water $(298-573$ K) as the working fluid. Copper—water heat pipes have a copper envelope, use water as the working fluid, and typically operate in the temperature range from 293 to 423 K. The working fluids have some limitations in the heat transfer rates of the heat pipe.

Development of nanofluids, i.e., fluids consisting of a conventional heat transfer base fluid with nanometer-sized oxide or metallic particles suspended within, offers the opportunities of increased heat transfer rates over conventional systems by more than 20%. Shukla et al. [88',89] reported a 30% increase in thermal conductivity by dispersing 0.1% copper particles in water. Significant improvement of heat transfer rates was found by dispersing copper, alumina, and silver colloid suspensions in water. Enhancement of thermal conductivity was also observed with copper oxide dispersed in water and ethylene glycol [90]. In addition to the heat transfer rates, the magnetic affinity of the solid particles in metallic suspensions allows for their manipulation by electromagnets, thereby eliminating the need for pumps and controls and may provide enhanced heat transfer. This has created interest in the application of a novel nanofluid-based actively controlled thermal management system for small satellite applications [91].

NASA has set a road map for the development of high-temperature heat pipes, which will be a solution for the high heat flux encountered during ascent and reentry of space vehicles [92]. The fluid with ultrafine suspended nanoparticles might be appropriate for satellite applications of heat pipes.

7.7 CONCLUDING REMARKS AND SUMMARY

This chapter presented a brief review of heat pipes in aerospace applications and also the basic principles of the operation of heat pipes. Thermal management in spacecraft is challenging because of the adverse environmental radiation. A variety of heat pipes for cryogenic and high-temperature applications have already been used in space applications. LHPs and CPLs with multiple evaporators and condensers have been found to be effective thermal management solutions for high-powered communication satellites. The macro and micro LHPs can play important roles in thermal solutions for small satellites. Heat pipes have emerged as appropriate and cost-effective responses to these challenges. Heat pipes used in thermal protection systems have been found superior to high-temperature materials, with the benefits of lightweight and a passive design. A wick having nanostructures, with a gradient along the length of the wick, can promote capillary fluid flow, and the use of nanofluids as working fluids can improve the heat transfer rates of heat pipes.

REFERENCES

[1] Shukla KN. Heat pipe for aerospace applications—an overview. J Electron Cool Therm Control 2015;5:1—14.
[2] Zemlianoy P, Combes C. Thermal control of space electronics. Electron Cool September 1, 1996. http://www.electronics-cooling.com/1996/09/thermal-control-of-space-electronics/.
[3] McIntosh R, Ollendorf S, Harwell W. The international heat pipe experiment. In: Proceedings of International Heat Pipe Conference, Bologna, April 1976; 1976. p. 589—92.
[4] Kirkpatrick JP, Brennan PJ. Long term performance of the advanced thermal control experiment. In: Proceedings of international heat pipe Conference, Bologna, April 1976; 1976. p. 629—46.
[5] Rankin JG. Integration and flight demonstration of a high capacity monogroove heat pipe radiator. In: Aiaa 19th Thermophysics Conference, Snowmass, AIAA paper No. 84—1716; 1984.
[6] Brown R, Gustafson E, Gisondo F, Harwell W. Performance evaluation of the Grumman prototype space Erectable radiator system. In: Aiaa paper No. 90—1766; 1990.
[7] Brown R, Kosson R, Ungar E. Design of the SHARE II Mono groove heat pipe. In: Proceedings of AIAA 26th Thermophysics Conference, Honolulu, AIAA paper No. 91—1359; 1991.

[8] Morgownik A, Savage C. Design aspect of a Deployable 10KW heat pipe radiators. In: Proceedings of the 6th international heat pipe Conference, Grenoble, 25-29 May 1987; 1987. p. 351—6.

[9] Amidieu M, Moscheti B, Taby M. Development of a space Deployable radiator using heat pipes. In: Proceedings of the 6th international heat pipe Conference, Grenoble, 25-29 May 1987; 1987. p. 380—5.

[10] Peck S, Fleischman G. Lightweight heat pipe Panels for space radiators. In: Proceedings of the 6th international heat pipe Conference, Grenoble, 25-29 May 1987; 1987. p. 36—367.

[11] Brennan PJ, Thienel L, Swanson T, Morgan M. Flight data for the cryogenic heat pipe (CRYOHP) experiment. AIAA; 1993. p. 93—2735.

[12] Schulze T, Sodtke C, Stephan P, Gambaryan-Rosisman T. Performance of heat pipe evaporation for space applications with combined Re-Entrant and Microgrooves. In: Proceedings of the 14th international heat pipe Conference, Florianopolis, 22-27 April 2007; 2007.

[13] Swanson TD. Thermal control techniques for the New Age of space exploration. In: Proceedings of the 14th international heat pipe Conference, Florianopolis, 22-27 April 2007; 2007.

[14] Barthelemy R, Jacobson D, Rabe D. Heat pipe mirrors for high power lasers. In: Proceedings of the 3rd international heat pipe Conference, Palo Alto, 22—24 May 1978, paper No. 78-391; 1978. http://dx.doi.org/10.2514/6.1978-391.

[15] Williams R. Investigation of a cryogenic thermal diode. In: Proceedings of the 3rd international heat pipe Conference, Palo Alto, 22—24 May 1978, paper No. 78-391; 1978. http://dx.doi.org/10.2514/6.1978-417.

[16] Thieme LG, Schreiber JG. Nasa GRC Stirling technology development overview. 2003. NASA/TM-2003—212454.

[17] Tarau C, Anderson WG. Sodium variable conductance heat pipe for radioisotope Stirling systems, design and experimental results. In: Proceedings of the 8th Annual international energy Conversion Engineering Conference, Nashville, 25-28 July 2010, AIAA 2010-6758; 2010.

[18] Glass DE, Camarda CJ, Mennigen MA, Sena JT, Reid RR. Fabrication and testing of a leading edge shaped heat pipe. J Spacecr Rockets 1999;36:921—3.

[19] Steeves OA, He MY, Kasen SD, Valdevit L, Wadley HNG. Feasibility of metallic structural heat pipes as Sharp leading edges for Hypersonic vehicles. ASME J Appl Mech 2009;76, 031014. http://dx.doi.org/10.1115/1.3086440.

[20] WWW.THERMACORE.COM, Press Release April 29, 2014.

[21] Chatterjee A, Wayner PC, Plawsky JL, Chao DF, Sicker RJ, Lorik T, et al. The constrained vapor bubble Fin heat pipe in microgravity. Ind Engn Chem Res 2011;50(15):8917—26. http://dx.doi.org/10.1021/ie102072m.

[22] Gaugler RS. Heat transfer devices. U.S. Patent 2,350,348. 1944.

[23] Grover GM, E Cotter T, E Erikson G. Structures of very high thermal conductivity. J Appl Phys 1964;35:1190—1.

[24] Bar-Cohen A, Kraus AD. Advances in thermal modeling of electronic components and systems, vol. 1. Washington, DC: Hemisphere Publishing Corporation; 1988. p. 283—336.

[25] Rohsenow WM, Hartnett JR, Cho YI. Handbook of heat transfer. 3rd ed. McGraw-Hill Co; 1998. p. 1—13 [Chapter 12] Heat Pipes.

[26] E Peterson G, Lu XJ, E Peng X, Wang BX. Analytical and experimental investigation of the Rewetting of circular channels with internal V-Grooves. Int J Heat Mass Transf 1992;35(11):3085—94.

[27] Peterson GP, Peng XF. Experimental investigation of capillary Induced Rewetting for a flat porous wicking structure. ASME J Energy Resour Technol 1993;115(1):62—70.

[28] Groll M, Supper W, Savage CJ. Shutdown characteristics of an axial-grooved liquid-Trap heat pipe thermal diode. J Spacecr 1982;19(2):173—8.

[29] Peterson GP. An Introduction to heat pipes: Modeling, testing and applications. Washington, DC: J. Wiley & Sons, Inc; 1994.

[30] Faghri A. Heat pipe Science and technology. Washington, DC: Taylor & Francis Publishing Company; 1995.

[31] Chi SW. Heat pipe theory and Practice. New York: McGraw-Hill Publishing Company; 1976.

[32] Ferrell JK, Alleavitch J. Vaporization heat transfer in capillary wick structures, preprint no. 6. In: Asme-aiche heat transfer Conf., Minneapolis, MN; 1969.

[33] Freggens RA. Experimental Determination of wick properties for heat pipe applications. In: Proc. 4th Intersoc. Energy Conversion Eng. Conf., Washington, DC; 1969. p. 888—97.

[34] Tien CL. Fluid Mechanics of heat pipes. Annu Rev Fluid Mech 1975:167—86.

[35] Dunn ED, Reay DA. Heat pipes. 3d ed. New York: Pergamon Press; 1982.

[36] Peterson GP, Bage B. Entrainment limitations in Thermosyphons and heat pipes. ASME J Energy Resour Technol 1991;113(3):147—54.

[37] Rice G, Fulford D. Influence of a fine mesh screen on entrainment in heat pipes. In: Proc. 6th Int. Heat pipe Conf., Grenoble, France; 1987. p. 168—72.

[38] Feldman KT. The heat pipe: theory, design and applications, technology application center. Albuquerque, NM: Univ. of New Mexico; 1976.

[39] Brennan PJ, Kroliczek EJ. Heat pipe design Handbook. Towson, MD: B&K Engineering, Inc; 1979.

[40] Gosse J. The thermo-physical properties of fluids on liquid-vapor equilibrium: an Aid to the Choice of working fluids for heat pipes. In: Proc. 6th Int. Heat pipe Conf., Grenoble, France; 1987. p. 17—21.

[41] Heine D, Groll M. Compatibility of Organic fluids with commercial structure materials for Use in heat pipes. In: Proc. 5th Int. Heat pipe Conf., Tsukuba, Japan; 1984. p. 170—4.

[42] Udell KS, Jennings JD. A composite porous heat pipe. In: Proc. 5th Int. Heat pipe Conf., Tsukuba, Japan; 1984. p. 41—7.

[43] Barantsevich VL, Barakove LV, Tribunskaja IA. Investigation of the heat pipe corrosion resistance and Service characteristics. In: Proc. 6th Int. Heat pipe Conf., Grenoble, France; 1987. p. 188—93.

[44] Basiulis A, Prager RC, Lamp TR. Compatibility and reliability of heat pipe materials. In: Proc. 2nd Int. Heat pipe Conf., Bologna, Italy; 1976. p. 357—72.

[45] Busse CA, Campanile A, Loens J. Hydrogen generation in water heat pipes at 250°C. In: First Int. Heat pipe Conf., Stuttgart, Germany, paper no. 4—2; October 1973.

[46] Zaho RD, Zhu YH, Liu DC. Experimental investigation of the compatibility of mild Carbon steel and water heat pipes. In: Proc. 6th Int. Heat pipe Conf., Grenoble, France; 1987. p. 200—4.

[47] Roesler S, Heine D, Groll M. Life testing with stainless steel/ammonia and aluminum/ammonia heat pipe. In: Proc. 6th Int. Heat pipe Conf., Grenoble, France; 1987. p. 211—6.

[48] Murakami M, Arai K. Statistical prediction of long-term reliability of copper water heat pipes from acceleration test data. In: Proc. 6th Int. Heat pipe Conf., Grenoble, France; 1987. p. 2194—9.

[49] Alario J, Brown R, Otterstadt E. Space constructable radiator prototype test Program. In: Aiaa paper No. 84—1793; 1984.

[50] Peterson GP. Heat removal Key to Shrinking avionics. no. 8, October. Aerospace America; 1987. p. 20—2.

[51] Basiulis A, Filler M. Operating characteristics and long life capabilities of Organic fluid heat pipes. In: Aiaa paper No. 71—408; 1971.

[52] Kreed H, Kroll M, Zimmermann P. Life test investigations with low temperature heat pipes. In: Proc. First Int. Heat pipe Conf., Stuttgart, Germany, paper no. 4—1; October 1973.

[53] Toth JE, Meyer GA. Heat pipes: is reliability an Issue?. In: Proc. Ieps Conf., Austin, TX; September 28, 1992.

[54] Baker E. Prediction of long term heat pipe performance from accelerated life tests. AIAA J September 1979;11(9).

[55] Faghri A. Heat pipes, review, opportunities and challenges. Front Heat Pipes (FHP) 2014;5:1—48. http://dx.doi.org/10.5098/fhp.5.1.

[56] Mochizuki M, Nguyen T, Mashiko K, Saito Y, Nguyen T, Wuttijumnong V. A review of heat pipe application including new opportunities. Front Heat Pipes (FHP) 2011;2, 01300.

[57] Mock PR, Marcus DB, Edelman EA. Communication technology satellite: a variable conductance heat pipe application. J Spacecr Rockets 1975;12:750—3.

[58] Marcus BD, Fleischman GL. Steady-state and transient performance of hot reservoir gas controlled heat pipes. In: Asme paper 70-HT/SPT-11; 1970.

[59] Edward DK, Marcus BD. Heat and mass transfer in the vicinity of the vapor-gas front in a gas- loaded heat pipe. J Heat Transf 1972;9:155—62. http://dx.doi.org/10.1115/1. 3449887.

[60] Rohani AR, Tien CL. Steady two-dimensional heat and mass transfer in the vapor-Gs region of a gas loaded heat pipe. J Heat Transf 1977;95:377—82. http://dx.doi.org/10. 1115/1.3450067.

[61] Sun KH, Tien CL. Thermal performance characteristics of heat pipe. Int J Heat Mass Transf 1975;18:363—80. http://dx.doi.org/10.1016/0017-9310(75)90026-5.

[62] Shukla KN. Transient response of a gas controlled heat pipe. AIAA J 1981;19:1063—70. http://dx.doi.org/10.2514/3.7842.

[63] Shukla KN, Sankara Rao K. Heat and mass transfer in the vapor gas region of a gas-loaded heat pipe. ZAMM—J Appl Math Mech 1983;63:575—80.

[64] Shukla KN. Thermal performance of a gas-loaded heat pipe. In: Proceedings of the second Asian Congress of fluid Mechanics, Beijing, 25—29 October 1983, 405. Science Press; 1983.

[65] Harley C, Faghri A. Transient two-dimensional gas-loaded heat pipe analysis. ASME, J Heat Transf 1994;116:716—23. http://dx.doi.org/10.2514/3.7842.

[66] Gray VH. The rotating heat pipe—a wickless, hollow shaft for transferring high heat fluxes. In: Proceedings of the ASME/AIChE heat transfer Conference, Minneapolis, August 3—6 1969; 1969. p. 1—5.

[67] Shukla KN, Solomon AB, Pillai BC. Experimental studies of rotating heat pipes. Heat Transf Asian Res 2009;38:475—84.

[68] Peterson GP, Compagua C. Review of cryogenic heat pipes in spacecraft applications. J Spacecr Rockets 1987;24:99—100.

[69] Charlton MC, Bowman WI. A mathematical model to predict the transient temperature profile of a cryogenic heat pipe during startup. J Spacecr Rockets 1994;31:914—6.

[70] Conto P, Ochterbeck JM, Montelli MBH. Analysis of supercritical startup of cryogenic heat pipes with parasitic heat loads. AIAA J Thermophys Heat Transf 2005;19:497—508.

[71] Bughy DC, Cepeda-Rizo J, Rodriguez JL. Thermal switching cryogenic heat pipe. In: Miller SD, Roes Jr RG, editors. International Cryocooler Conference—Cryocoolers, 16. Boulder: ICC Press; 2011. p. 557—66.

[72] Wu XP, Mochizuki M, Nguyen T, Saito Y, Wuttijumnong V, Ghisoiu H, et al. Low profile high performance vapor chamber heat sinks for cooling high density blade servers. In: 23rd Annual IEEE Semi-Therm Symposium; 2007. p. 174—8.

[73] Xiao B, Faghri A. Three dimensional thermal fluid analysis of flat heat pipes. Int J Heat Mass Transf 2008;51:3113—26. http://dx.doi.org/10.1016/j.ijheatmasstransfer.2007.08.023.

[74] Shukla KN, Solomon AB, Pillai BC. Thermal performance of vapor chamber with nanofluids. Front Heat Pipes (FHP) 2012;3, 033004.

[75] Maydanik YF. Loop heat pipes. Appl Therm Eng 2005;25:635—57. http://dx.doi.org/10.1016/j.applthermaleng.2004.07.010.

[76] Ku J. Operating characteristics of loop heat pipes. In: Proceedings of the 29th international Conference on environmental system, Denver, 1—15 July 1999, paper No. 1999-01-2007; 1999.

[77] Chemyshea MA, Maydanik YF, Ochterbeck JM. Heat transfer investigation in evaporator of loop heat pipe during startup. AIAA J Thermophys Heat Transf 2008;22:617—22.

[78] Chemyshea MA, Maydanik YF. Heat and mass transfer in evaporator of loop heat pipe. AIAA J Thermophys Heat Transf 2009;23:725—31.

[79] Shukla KN. Thermo fluid dynamics of loop heat pipe operation. Int Commun Heat Mass Transf 2008;35:916—20. http://dx.doi.org/10.1016/j.icheatmasstransfer.2008.04.020.

[80] Ku J, Ottenstein L, Douglas L, Pauken M, Nirur G. Miniature loop heat pipe with multi evaporators for thermal control of small spacecraft. In: The American Institute of Aeronautics and Astronautics, AIAA paper No. 183; 2010.

[81] Ku J, Paiva K, Mantelli M. Loop heat pipe transient behavior using heat source temperature for set point control with thermoelectric converter on reservoir. NASA—Goddard Space Flight Center; September 2011.

[82] Okutani S, Nagano H, Okazaki S, Ogawe H, Nagai H. Principles and prospects for micro heat pipes. J Electron Cool Therm Control 2014;4, 43507.

[83] Cotter TP. Principles and prospects for micro heat pipes. Proceedings of the 5th international heat pipe conference, Tsukuba, 14—18 May 1984; 1984. pp. 328—335.

[84] Peterson GP. Overview of micro heat pipe research and development. Appl Mech Rev 1992;45:175—89. http://dx.doi.org/10.1115/1.311975.

[85] Hopkins R, Faghri A, Khrustalev D. Flat miniature heat pipes with micro capillary grooves. J Heat Transf 1999;121:102—9. http://dx.doi.org/10.1115/1.2825922.

[86] Shukla KN. Heat transfer limitation of a micro heat pipe. ASME J Electron Packag 2009;131, 024502. http://dx.doi.org/10.1115/1.3103970.

[87] Gerner FM, Longtin JP, Henderson HT, Hsieh WM, Ramdas P, Chang WS. Flow limitations in micro heat pipes. In: Proceedings of the 28th ASME National Heat Transfer Conference, San Diego, 9—12 August 1992; 1992. p. 99—104.

[88] Shukla KN, Solomon AB, Pillai BC, Ibrahim M. Thermal performance of cylindrical heat pipe using nanofluids. AIAA J Thermophys Heat Transf 2010;24:796—802. http://dx.doi.org/10.2514/1.48749.

[89] Shukla KN, Solomon AB, Pillai BC, Jacob Ruba Singh B, Kumar SS. Thermal performance of heat pipe with suspended nano-particles. Heat Mass Transf 2012;48:1913—20. http://dx.doi.org/10.1007/s00231-012-1028-4.

[90] Wang X, Xu X, Choi SUS. Thermal conductivity of nanofluid mixture. AIAA J Thermophys Heat Transf 1999;13:474—80. http://dx.doi.org/10.2514/2.6486.

[91] Li YY, L.C LV, Liu ZH. Influence of nanofluids on the operation characteristics of small capillary pumped loop. Energy Convers Manag 2010;51:2312—20. http://dx.doi.org/10.1016/j.enconman.2010.04.004.

[92] Hill SA, Kostyk C, Motil B, Notardonato W, Rickman S, Swanson T. Thermal management systems roadmap technology area-14. NASA—2012; 2012.

CHAPTER 8

Fuel Cells

8.1 INTRODUCTION

A fuel cell is a very effective power source. It is commonly defined as an electrochemical device that converts the supplied fuel to electric energy and heat continuously as long as reactants are supplied to its electrodes. It has no moving parts, works quietly, and emits only water vapor. The sound created may come from an air compressor or a cooling fan, depending on the type of fuel cell and how it is operating or for what purpose it is used.

In principle a fuel cell is working as a battery but it continues to operate as long as fuel is supplied. Basically the fuel cell is like a thin-layered sandwich consisting of two electrodes, an anode and a cathode, on each side of an electrolyte. Such a simple cell generates only a small amount of power and to generate a sufficient electric current a number of unit cells need to be assembled in the so-called stack. Ideally hydrogen is supplied on the anode side and oxygen or air on the cathode side. However, hydrogen is almost not freely available in the atmosphere because it attempts to find oxygen atoms for marriage and creates water. This attraction ability is taken care of in a fuel cell.

Fig. 8.1 shows a principle sketch of the so-called proton exchange membrane fuel cell (a.k.a. polymer electrolyte membrane fuel cell, PEMFC) in which the electrolyte is a polymer. The hydrogen is supplied on the anode side and it tries to reach the oxygen on the cathode side. In doing so, it achieves assistance from a catalytic material such as platinum. On the catalytic material the hydrogen atoms are split up into protons and electrons, which find different ways to join with the oxygen. The protons cross the electrolyte, whereas the electrons need to take a way through the electric circuit to reach the cathode. In the process, electricity and heat are generated. As the protons are joining with the oxygen, water is created. The noble metal platinum is a key factor in PEMFCs and it is regarded as a stable catalyst. However, it is a limited resource and research work is underway to find a substitute for it.

Heat Transfer in Aerospace Applications
ISBN 978-0-12-809760-1
http://dx.doi.org/10.1016/B978-0-12-809760-1.00008-9

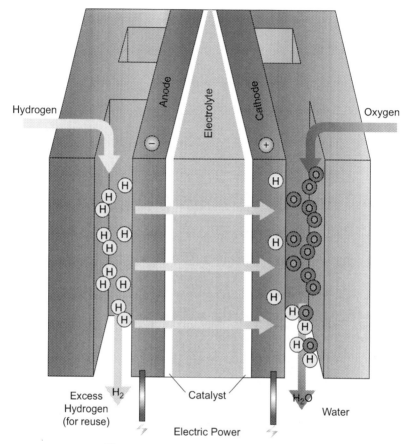

Figure 8.1 Principle sketch of a PEMFC.

8.2 TYPES OF FUEL CELLS

There are several types of fuel cells and each one has advantages and disadvantages. Table 8.1 describes a brief summary of some types of fuel cells and their characteristic features, operating temperatures, and areas of application. More detailed descriptions are presented in the following sections.

8.2.1 Proton Exchange Membrane Fuel Cells or Polymer Electrolyte Fuel Cells (PEFCs)

This fuel cell operates at a relatively low temperature, about 80 C. The energy conversion efficiency, i.e., conversion of the hydrogen energy to electricity is in the order of 50–60%. The electrolyte is a solid polymer in

Table 8.1 Principle Operation of Some Fuel Cell Types

Fuel Cell Type	Mobile Ion	Operating Temperature, °C	Applications
PEMFC and PEFC	H^+	30–100	Vehicles, mobile equipment, low-power CHP systems
DMFC	H^+	20–90	Portable electronic systems with low power, long operating times
PAFC	H^+	~200	Large numbers of 200-kW CHP systems
AFC	OH^-	50–200	Space vehicles
MCFC	CO_3^{2-}	~650	Medium- to large-scale CHP systems
SOFC	O^{2-}	500–1000	All sizes of CHP systems

CHP, combined heat and power; *PEFC*, polymer electrolyte fuel cell.

which protons are mobile. This is the fuel cell type being promoted and introduced as the power source for ground vehicles. It has a good start-up behavior at low temperatures and is small and compact. It is also considered as the power source in portable units, as well as in large stationary power plants. An advantage is that the electrolyte is a solid phase in form of a thin membrane of a polymer. The surrounding catalyst is platinum. This cell is sensitive to the purity of the hydrogen fuel. PEMFCs were used on the first manned spacecraft. The platinum catalyst is very costly, but nickel-tin catalysts have been discovered. As an effect, fuel cells can be a possible substitute for batteries in spacecraft.

8.2.2 Alkaline Fuel Cells

This has been used in space vehicles for generation of electricity and drinkable water for astronauts. The conversion efficiency is about 70%, and the operating temperature is in the range of 150–200°C. The electrolyte mainly consists of potassium hydroxide (KOH). Some interests have also been shown for their use in ground vehicles. An advantage of this fuel cell type is that it does not depend on platinum as the catalyst. Carbon dioxide may affect the conversion and it is generally required that the supplied air and fuel must be free from CO_2, otherwise pure oxygen and hydrogen must be used. This fuel cell type was used on the Apollo and Shuttle Orbiter craft.

8.2.3 Phosphoric Acid Fuel Cells (PAFCs)

This fuel cell type has been commercially in operation for a while. It has found applications in, e.g., hospitals, hotels, offices, airports, and schools.

The conversion efficiency is relatively low, about 40–50%. The operating temperature is in the range of 150–200°C. Both the electricity and steam can be taken care of. The electrolyte is an acid of phosphorus. This fuel cell type resists pollution created by the hydrogen fuel particularly at high temperatures. At low temperature, carbon monoxide may damage the platinum layer on the catalyst. Units of this fuel cell type are often big and heavy but the technology is regarded as mature.

8.2.4 Solid Oxide Fuel Cells

This fuel cell type is regarded as a candidate for large units even at remote locations. It has also found application as auxiliary power units (APUs) even in vehicles. The electrolyte is in a solid state and commonly made in a hard ceramic material based on zirconium oxide. The operating temperature is higher, around 1000°C; hence, high reaction rates can be achieved without expensive catalysts and gases, e.g., natural gas, can be used directly or internally reformed without the need of a separate unit. The conversion efficiency is about 60%. It may also be integrated with a steam turbine to generate additional electricity. It does not necessarily require pure hydrogen as the fuel.

8.2.5 Molten Carbonate Fuel Cells (MCFCs)

The electrolyte consists of a melt of carbonates of lithium, sodium, and potassium. The conversion efficiency is about 60%, and the operating temperature is about 650°C. It needs carbon dioxide in the air to work. The high operating temperature means that a good reaction rate is achieved by using a relatively cheap catalyst, namely, nickel. The nickel catalyst also forms the basis of the electrode. Its simplicity is partly offset by the nature of the electrolyte, which is a hot and corrosive mixture of lithium, potassium, and sodium carbonates. Similar to other high-temperature fuel cells, it can be combined with a steam turbine to generate additional electricity. It can be operated with other hydrogen carriers rather than pure hydrogen. The high-temperature operation, as for solid oxide fuel cells (SOFCs), may have a severe effect on the cell components.

8.2.6 Direct Methanol Fuel Cells (DMFCs)

This is a variant of the PEMFC but the methanol reacts directly in the anode rather than in a separate reformer upstream the fuel cell. The conversion efficiency is about 30–40%, and the operating temperature is in the range of 50–100°C. The loss in the reformer is eliminated. It may find applications in portable computers and mobile phones.

8.2.7 Reversible Fuel Cells

A cell that can be operated both as a fuel cell and as an electrolyzer is called a reversible fuel cell. In case of surplus of electricity available from the wind or the sun, the cell can be operated to generate hydrogen by electrolysis of water. As it operates in the fuel cell mode, it uses hydrogen as the fuel and generates electricity.

8.2.8 Proton Ceramic Fuel Cells

This is a fuel cell that is under development. The idea is that the ceramic electrolyte should be able to conduct protons. The operating temperature is about 700°C, and hydrogen is oxidized directly at the anode. A reformer is not needed. This fuel cell type is believed to be able to combine the advantages of the high-temperature fuel cells and the PEMFCs because of the use of a ceramic electrolyte.

Several books on fuels cells are available, some are more general and others are specific to a certain fuel cell type, see, e.g., [1–6].

8.3 BASIC TRANSPORT PROCESSES AND OPERATION OF A FUEL CELL

The fuel cells discussed so far function in the same manner. At the anode a fuel, commonly hydrogen, produces free electrons, and at the cathode, oxygen is reduced to oxide species. Depending on the electrolyte, either protons or oxide ions are transported through the ion-conducting, but electronically insulating, electrolyte to combine with oxide or protons to produce water and electric power. The reactions at the anode and cathode sides must proceed continuously and then the electrons produced at the anode have to pass an electric circuit to the cathode. Ions need to migrate through the electrolyte. The electrolyte is only permitted to transport ions not electrons. However, the reaction characteristics at the anode and cathode are different for different fuel cells but the overall reaction is the same.

8.3.1 Electrochemical Kinetics

Studies of electrochemical kinetics is important for design and operation of fuel cells. The electron transfer rate at the electrodes or the current produced by the fuel cell depends on the rate of electrochemical reaction. The processes governing the electrode reaction rates are the mass transfer between the bulk solution and the electrode surface, the electron transfer at the electrode, and the chemical reactions involving electron transfer.

8.3.2 Heat and Mass Transfer

A number of transport processes occur in a fuel cell. The reactant gases flow through the gas flow channels and reactant gas species are transported from the gas flow channels and through the porous electrodes. Ions are transported through the membranes or the electrolyte and electrons are transported through electrodes and interconnects. Fig. 8.2 gives a principle illustration.

Poor transport of heat and mass contributes to the loss of fuel cell performance. Charge transport contributes to ohmic losses and the mass transfer of the reactant gases impacts the mass transfer losses.

8.3.3 Charge and Water Transport

Electrons and ions are produced and consumed in two electrochemical reactions at the anode—electrolyte and cathode—electrolyte interfaces. The electrons are transported through the el and interconnect to the external electric circuit. Ions are transported through the electrolyte from the

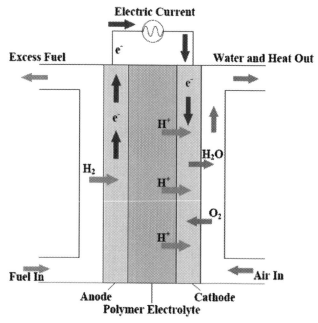

Figure 8.2 Fluid flow and heat and mass transfer in a fuel cell of three layers.

electrode where they are produced to the electrode where they are consumed. Ohmic voltage losses are caused by the resistances to the motion of ions through the electrolyte as well as that of electrons through the electrodes, interconnect materials, and contact interfaces. Ion transport is also important for the transport of water in PEMFCs.

8.4 AEROSPACE APPLICATIONS

Use of hydrogen in flying vehicles has a long history. It started with balloons and then continued with airships. However, after the Hindenburg accident, use of hydrogen as power source became less popular.

The fuel cells made a comeback for use in the aerospace applications. The first practical application in space was the Gemini project in the 1960s and the fuel cell used was a polymer electrolyte fuel cell (PEFC). The alkaline fuel cells (AFCs) were used in, e.g., the Apollo vehicles for the lunar trips. With the fuel cells, it was possible to generate electricity even when the vehicle and its solar cells were not exposed to the Sun's radiation. The restulting product could be used as drinkable water. Nowadays, PEFCs or PEMFCs are considered again for aerospace applications.

Fuel cells are more efficient than secondary batteries. Theoretically a fuel cell is able to deliver 500 kWh/kg of hydrogen plus oxygen. The currently best lithium batteries can deliver about 120 kWh/kg. Fuel cells can convert fuel to electric power with an efficiency of over 80%. Typically a diesel combustion engine cannot provide an efficiency better than 40% at its optimum speed and load. For space applications with long discharge times, the mass of the fuel cell and other process units is less compared to that of stored fuel, oxidant, and tankage. For applications that run for a short time at moderate power levels, the mass of the fuel cell and other process units is significant, and reduces their competitiveness compared to batteries.

The fuel cell for space applications has specific system requirements and different operating conditions and designs because of the isolated low-gravity environment in space when compared to ground operation.

For ground applications at atmospheric pressure, air can be supplied as the cathode gas and hydrogen as the anode gas. The fuel cell reaction produces water and electricity. The water must be evaporized by compressed air because if water remains in the path in the separator, it may block the flow of the reactant air. However, proton conduction in the membrane requires water, and the air is humidified before being supplied to the fuel cell. Obviously, the fuel cell at ground applications requires humidification of the supplied gas and dehydration of the separator to balance the complicated humidity conditions inside the fuel cell.

For space applications, it is very important to prepare a simple system. High power or energy density and air-independence are required. As the spacecraft is very isolated in the Earth's orbit, all reactant materials must be carried inside the spacecraft. In order to minimize the weight, pure anode and cathode materials must be used and should be completely consumed. The produced water must be collected. This has created a renewed interest on the use of PEMFCs in these applications.

Another type of fuel cell using novel fuel and oxidizer has also been investigated for space power systems. This is because of the revived interest on using hydrogen peroxide in aerospace power applications. Hydrogen peroxide (H_2O_2) is used directly at the cathode. The fuel on the anode side is hydrogen gas and an aqueous $NaBH_4$ solution. It has been found that the direct utilization of H_2O_2 and $NaBH_4$ at the electrodes resulted in about 30% higher voltage output than that of a regular H_2-O_2 fuel cell. Both $NaBH_4$ and H_2O_2 are in aqueous form and the combination has some operational advantages. A Nafion membrane is used as the electrolyte. The catalysts are commonly platinum based. The carbon substrate used is here called reactant diffusion layer instead of gas diffusion layer because the peroxide reactant is in the liquid phase. The design is compact and it is ideal for space applications where a high-energy-density fuel is required and air is not available.

Hydrogen peroxide is commonly used in rocket propulsion and air-independent power systems. It has also found utilization in underwater power systems. It is a powerful oxidizer and is safe. After it is decomposed, it gives only oxygen and water and thus creates no environmental problems.

The benefits of a direct hydrogen peroxide fuel cell over fuel cells using gaseous oxygen are higher current density because of larger oxidizer mass density, single-phase transport on the cathode side of the fuel cell increases the reaction rate, and elimination of the oxygen reduction overpotential problem.

REFERENCES

[1] Revankar S, Majumdar P. Fuel cells—principles, design and analysis. New York: CRC Press; 2014.
[2] Singhal SC, Kendall K. High temperature solid oxide fuel cells-fundamentals, design and applications. Oxford, UK: Elsevier; 2004.

[3] Hodgers G, editor. Fuel cell technology handbook. New York: CRC Press; 2003.

[4] Barbir F. PEM fuel cells, theory and practice. New York: Academic Press; 2012.

[5] Sunden B, Faghri M, editors. Transport phenomena in fuel cells. Southampton, UK: WIT Press; 2005.

[6] Larmine J, Dicks A. Fuel cell systems explained. London, UK: J. Wiley & Sons; 2000.

CHAPTER 9

Microgravity Heat Transfer

9.1 INTRODUCTION

Microgravity is an environment where effects of gravity are small. The topic of microgravity is very relevant to aerospace applications. To the public, it is known as a weightless state. However, heat transfer under microgravity differs significantly from that under normal gravity. Heat conduction in solids and liquids is not affected by gravity, but in gases, it is indirectly reduced in low gravity because the gas density is reduced. Thermal radiation is not affected by gravity. However, phase change processes such as evaporation, boiling, and condensation, as well as two-phase forced convection and phase-change heat transfer are affected by the gravity. As the buoyancy forces become insignificant, other matters become important, such as capillary forces, viscous forces, and electro-magnetic forces. Topics to be considered in this chapter are capillary and two-phase flow in microgravity, bubble dynamics, boiling heat transfer, droplets and bubbles, and solidification. The thermal stratification and pressurization in a cryogenic tank partially filled with liquid hydrogen (LH$_2$) under reduced gravity will be illustrated. Microgravity can be created in two ways. Because the gravitational attraction diminishes with distance, a microgravity environment can be created by traveling away from the Earth. However, to create a microgravity (say 10^{-6} g) one needs to travel into space a distance of about $6.4 \cdot 10^6$ km from the Earth, i.e., about 17 times farther away than the moon. This is not very practical except for spacecraft. A more practical way to create a microgravity environment is by using free fall. At low Earth orbit the gravity force is about 89% of the sea level value but one does not feel it in orbit because one is in a state of perpetual free fall. In orbit, one flies fast and high enough not to fall and hit the Earth. The centripetal force from the circular motion is equal and opposite to the force of gravity.

9.2 SOLIDIFICATION IN MICROGRAVITY

Gravity induces significant disturbances in gaseous and liquid systems. These disturbances are manifested in the form of sedimentation, thermal

Heat Transfer in Aerospace Applications
ISBN 978-0-12-809760-1
http://dx.doi.org/10.1016/B978-0-12-809760-1.00009-0

convection, and hydrostatic pressure. All these phenomena have an impact on the solidification characteristics of liquid metals and alloys and semi-conducting materials. By carrying out a solidification process in micro-gravity environment, the gravity-induced disturbances can be virtually eliminated and the solid structures produced are unique in uniformity of composition and properties, i.e., a homogeneous microstructure is achieved. On a space vehicle in orbit, a low-gravity environment is created and then fundamental investigations of crystal growth can be carried out. The convection due to gravity is eliminated or at least reduced to a level at which crystal growth is mainly diffusion controlled. Accordingly, many crystal growth experiments have been performed in the microgravity environment of a spacecraft. However, in such environments, there still might be variations in the magnitude and direction of the gravity. Such variations may have significant effects on the solute distribution in the melt.

A brief review of some research activities in microgravity that are of relevance to solidification in relation to material sciences was presented [1]. Some information on processing of eutectics in microgravity, interface undercooling, solidification velocity and nucleation, isothermal dendritic growth, macrosegregation in alloys, and others was also presented.

Permanent magnets of Nd-Fe-B alloys have been used in various applications because of their superior energy products. Commonly they are prepared by either sintering techniques or rapid solidification processing techniques. However, attempts have been made to solidify the melts under microgravity conditions to obtain a homogeneous microstructure. In one study [2], Nd-Fe-B alloy ingots were melted and superheated and then quenched under a microgravity of 10^{-5} g_0. It was found that the resulting alloy ingots consisted mostly of equiaxed $Nd_2Fe_{14}B$ grains, which were similar to those obtained under normal gravity conditions. However, the grain size distribution under microgravity was smaller than that under normal gravity. In the grain boundary regions, α-Fe- and Nd-rich phases were observed.

In casting, commonly two types of dendrites appear: columnar and equiaxed. The transition of a macrostructure from columnar to equiaxed is called the columnar to equiaxed transition (CET). This transition is important, as it strongly affects the properties of the cast components. A limited number of solidification experiments under microgravity conditions have been conducted to study the CET. The European Space Agency (ESA) has coordinated a specific project and the Materials Science

Laboratory (MSL) onboard the International Space Station (ISS) has been used as the microgravity platform. A numerical simulation approach to a directional solidification experiment was conducted [3]. The numerical approach combined a front-tracking method of the columnar growth and an equiaxed volume averaging method. The combined model was based on the solution of the energy equation, with a latent heat source term. The predicted CET position showed good agreement with the experimental results.

Murphy et al. [4] experimentally studied the equiaxed solidification of Al-Cu alloys under microgravity and hypergravity. The solidification was controlled such that nucleation occurred coincidently with the onset of microgravity. This allowed for the effects of microgravity on equiaxed nucleation and initial growth.

A numerical investigation of the importance of the Marangoni effect on solidification under microgravity conditions was presented in Ref. [5]. The mathematic formulation was based on the enthalpy method. The effect of convective boundary condition above the free surface and the presence of argon were considered. The influence of different liquid and solid conductivities was studied. It was found that the Marangoni convection and the presence of argon affected the solidification process. The argon layer and the different liquid and solid conductivities resulted in a significant curvature of the interface.

A successful result of microgravity research is the development of bulk metallic glasses. These are a class of noncrystalline metal alloys. Compared to conventional crystalline metal alloys such as steel, aluminum, and titanium, the bulk metallic glasses are not well-defined materials, as their mechanical properties are difficult to assess without destructive testing. The bulk metallic glasses are undercooled liquids that have been captured as an amorphous solid by rapid cooling to below the glass transition temperature, without intervening crystallization. The bulk metallic glasses may find an application area in the space environment in structural components, such as debris shields, paneling, cellular structures, and mirrors. Further details can be found in Ref. [6].

In an experimental investigation of lead tin (PbSn) and lead telluride (PbTe) eutectic alloys under microgravity [7], it was concluded that for the PbSn eutectic alloy, no dendritic structures were formed and a uniform solute distribution appeared in contrast to normal gravity conditions. For the PbTe eutectic alloy the microgravity condition did not cause any changes compared to normal gravity conditions.

9.3 GRAVITY EFFECTS ON SINGLE-PHASE CONVECTION

Natural or free convection over small objects is of interest in many engineering applications. Because of the influence of gravity, natural convection on the ground will be different from that on orbit. A study [8] on facility drawer racks accommodating electronic units showed that for air convection the influence of gravity was negligible. In the same study, a manned cabin providing comfortable condition for astronauts by cold airflow was considered. It was found that the value of gravity was important, especially for vertical cases.

The investigation by Kostoglou et al. [9] concerned experiments in water and glycerol, in which the temperature evolution of a submillimeter spherical object with a temperature-dependent heat source was recorded during varying microgravity conditions. It was found that for low-gravity conditions the heat transfer mechanism is only conduction.

9.4 CONDENSATION UNDER MICROGRAVITY

In the Earth-based condensers, the condensate is drained commonly by the gravitational force. In the absence of gravity, such as in space applications, the condensate can be removed by suction, surface rotation, the shear stress due to vapor flow, capillary pumping, or a mechanical wiper. The literature on condensation in reduced gravity or microgravity environment is relatively sparse. A review was presented in Ref. [10]. Thus existing correlations for the condensation pressure drop and heat transfer are based on experiments carried out at normal gravity on the Earth. In an analysis of the condensation heat transfer process in a microgravity environment, two mechanisms for condensate removal were considered [11]. The first one concerned condensation on a flat porous plate, with condensate removal by suction. The second one predicted the heat transfer coefficient of a condensing annular flow, in which the condensate film was driven by vapor shear. It was concluded that both suction and vapor shear can effectively drain the condensate, thereby ensuring continuous operation in microgravity. A method to study condensation heat transfer inside a tube in a microgravity environment was introduced [12]. The method assumed laminar flow in the condensate film and that annular flow prevailed. The gravity was varied from zero to normal. On the Earth, gravity not only removes the condensate but also separates the vapor from the liquid. In space, gravity has no effect. In microgravity, the vapor closest to the tube wall condenses first. By surface tension the condensate will be absorbed and will form a symmetric annular film. The vapor velocity is higher than the liquid velocity, so obviously there is an interface between the two phases. Thus the simplest way to remove the condensate is by vapor shear. The results of the calculations showed that the heat transfer coefficient for normal gravity is higher than that for zero gravity because the condensate film is thinner and accordingly the thermal resistance is smaller in normal gravity. The fluid used was R-11 (trichlorofluoromethane). An interesting analysis of models and experimental data for condensing two-phase flow regimes, pressure gradients, and local heat transfer coefficient in the Earth normal gravity and in microgravity conditions inside a horizontal tube was presented by Mishkinis and Ochterbeck [13]. The media used were ammonia, propylene, and R-134a (1,1,1,2-tetrafluoroethane). It was found that (1) the length of condensation in microgravity at certain conditions can be larger than that in normal gravity and (2) the pressure drops in the tube for microgravity and normal gravity are not significantly different. For the heat transfer in annular flow, no principle physical difference between normal gravity and microgravity cases was found. Lee et al. [14] presented an experimental and theoretical investigation of annular flow condensation of FC-72 (perfluorohexane) in microgravity. The experiments were carried out in a parabolic flight. The flow behavior of the condensate film was shown to be sensitive mostly to the mass velocity of FC-72. Low mass velocities yielded laminar flow with a smooth interface, whereas for high mass velocities, turbulent flow with appreciable interfacial waviness prevailed. A number of tests performed at microgravity, lunar gravity, and Martian gravity proved that the influence of gravity was most pronounced at low mass velocities. This was manifested by a circumferential uniformity for microgravity versus an appreciable thickening along one side of the condensation tube for lunar and Martian conditions. At high mass velocities, the thickening did not appear for lunar and Martian gravities because of the increased vapor shear on the film interface, which proved that a high mass velocity is an effective means to eliminate the influence of gravity in space missions. For microgravity conditions, the condensation heat transfer coefficient was highest near the inlet (thin film and laminar flow) and then decreased along the tube length but finally increased again for high mass velocities as turbulence and waviness appeared. In addition a control-volume-based model was proposed to predict the condensation process, and the model accounted for dampening of the turbulent fluctuations near the film interface. Reasonable agreement with the experiments was found.

Transient laminar film condensation along a vertical flat plate under sinusoidal time-varying gravity has significant relevance in space in the condensing radiator for spacecraft thermal management system. In Ref. [15] an extension of the well-known Nusselt film theory (see, e.g., Ref. [16]) was presented. The boundary layer thickness and accordingly the heat transfer coefficient were found to oscillate with time.

9.5 BOILING/EVAPORATION IN MICROGRAVITY

The limited knowledge of the influence of gravity on multiphase flow and phase-change heat transfer is known to be one of the primary obstacles for reliable development and design of space-based equipment and processes such as heat exchange, cryogenic fuel storage, and transportation and cooling of electronics. If gravity effects on two-phase flows can be quantified, the size and weight of such systems can be reduced, which means that the costs of launching the components and space-based systems are lowered. Experimental investigations have shown that stable subcooled boiling on flat plates in microgravity environments is possible, although usually with reductions in the heat transfer coefficient of up to 50% compared with the Earth gravity values. Only a very limited amount of data is available for the local heat transfer rates under and around the bubbles as they grow and depart from the surface. Such data can provide necessary information concerning the relevant wall heat transfer mechanisms during the bubble departure cycle by clarifying when and where in the cycle large amounts of heat are removed.

Microgravity effects on the critical heat flux is another important area that must be addressed if boiling is to be used reliably as a heat removal mechanism, and it is another area with very limited quantitative information available. By performing tests in microgravity as well as lunar and Martian gravities, it is possible to assess the effect of buoyancy on the overall boiling process and determine the relative magnitude of effects with regard to other forces and phenomena such as the Marangoni forces, liquid momentum forces, and microlayer evaporation.

Microgravity environments can be produced by drop towers, parabolic aircraft, sounding rockets, and space flight. In drop towers, many drops can be generated per day and a high-quality microgravity (10^{-4} g) can be achieved but the microgravity time is limited to approximately 10 s. With the parabolic aircraft, many low-gravity periods per flight can be achieved but the microgravity quality is low (0.02 g). With sounding rockets, high-quality microgravity can be achieved (10^{-6} g) and can last for some minutes per flight. However, it is an expensive method. The space flight is certainly a good method with microgravities less than 10^{-6} g and lasting for longer periods, but it is also a very expensive method.

Microgravity pool boiling of hydrogen is of interest in many space applications and for the deep space exploration to moon and Mars. An improved understanding of hydrogen boiling in microgravity is of importance for reliable design and operation of space systems. A huge amount of pool boiling heat transfer investigations have been carried out under normal gravity conditions and a huge data base has been accumulated. On the other hand, the literature on low-gravity pool boiling is small because of high cost, hardware complexity, and short duration of the experimental systems, in particular for cryogenic liquids. The relationship between heat transfer at normal gravity and at microgravity is important to reveal. In an investigation [17] an existing gravity scaling analysis was assessed by considering pool boiling data of hydrogen under various gravities. It was found that the scaling analysis was acceptable in predicting the nucleate boiling heat transfer of hydrogen in low-gravity conditions. Another scaling analysis (power law relation) was found to be applicable in the film boiling regime. It was also found that when considering the boiling curves at normal and microgravity conditions, the latter yielded lower heat flux throughout the overall boiling regime and the influence of microgravity on the heat flux was most significant in the film boiling regime. In another study [18], under microgravity, the growth characteristics of a single bubble in a rectangular pool was investigated numerically by using the volume of fluid (VOF) multiphase model. The results showed that the temperature and the flow fields around the bubble are significantly different from those at normal gravity. The temperature profile at the two-phase interphase was not uniform, and the Marangoni effects were obvious. The detachment of the bubble was also affected by the gravity level. The surface tension caused by the Marangoni effect was found to be more significant at microgravity. It was also found that the bubble growth was more complex. At low gravity the bubble growth rate became higher. In addition the bubble diameter is affected by the gravity level and the fluctuating amplitude of the heat transfer coefficient increased with increasing microgravity. Flow boiling under various gravity conditions was investigated [19]. The experiments at low gravity were carried out with parabolic flights. The effects of mass flux, heat flux, and tube diameter on flow boiling at microgravity were studied. Compared to the result for terrestrial gravity, it was found that both heat transfer enhancement and deterioration could appear in microgravity conditions. The difference in bubble size was assumed to be a reason, as, in general, the bubble size in flow boiling is affected by the gravity level and the ratio of inertia forces to buoyancy forces. Heat transfer enhancement at low gravity was found for situations in which the flow pattern is bubbly in normal gravity and intermittent in low gravity. The results were also presented in a flow boiling gravity influence map, which might be a useful tool in designing boiling systems for space applications.

9.6 MICROGRAVITY EFFECTS IN CRYOGENIC TANKS

In Chapter 5, the behavior in cryogenic tanks for LH_2 under normal gravity was presented. In this section, complementary information is provided in terms of the effect of reduced gravity. Also the impact on the phase change process is discussed.

Future operations of many fluid, thermal, and power systems and their ability to store, transfer, and manage a variety of single- or multiphase fluids in reduced gravity environment are of great importance. For many of these systems, cryogenic conditions prevail. Cryogenic liquids such as LH_2 and liquid oxygen play an integral role in aerospace missions because of their high efficient thrust and nonpolluting waste. The cryogens are usually stored at a very low temperature and the storage tanks are equipped with high-quality insulation. However, heat leakage still occurs by conduction to the support structure and the pipelines that connect the tank to other devices. The heat will be carried to the liquid—vapor interface by conduction and natural convection, causing vaporization, which, in a closed tank, will result in a pressure rise. Cryogenic vaporization, caused by heat leakage into the tank from the surrounding environment, is one of the main causes of mass loss and leads to self-pressurization of the storage tanks. Controlling the self-pressurization and the pressure, as well as the thermal stratification, in cryogenic storage tanks is important for, e.g., space missions. Available publications on self-pressurization and stratification of cryogenic tanks mainly focus on the convection and surface evaporation influences. Because large superheats increase the likelihood of evaporation in the liquid, evaporation and its effect on vapor pressure under microgravity needs consideration. In this section the results from a numerical investigation [20] are highlighted. Effects of reduced gravity, the contact angle of the vapor bubble, and surface tension are presented. The computations were carried out by using the CFD software package, ANSYS

FLUENT, but an in-house developed code to calculate the source term associated with the phase change was implemented. A coupled level set and volume of fluid (CLSVOF) method was used to solve a single set of conservation equations for the whole domain and the interface between the two phases was tracked or captured. In addition a heat and mass transfer model was implemented to simulate the processes involving evaporation or condensation. Results show that small tiny vapor regions caused by the evaporation process change the pressure rise. Vortices are observed because of the vapor dynamics.

In this study a further development of the research works [21,22] is presented. In particular, evaporation and its effect on vapor pressure under microgravity are studied in a closed partially filled LH_2 tank by employing the CLSVOF method [23] and the phase change effect is also considered. The phenomenon is investigated in a partially filled LH_2 tank for different values of the gravity acceleration and the effects of contact angle and surface tension are studied.

An axisymmetric cylindrical tank partially filled with LH_2 is considered, as shown in Fig. 9.1. The diameter is 0.5 m and the height is 1 m. Only hydrogen vapor is considered to be present in the vapor space (ullage space). The VOF method [24], which is a kind of an Eulerian method, was used. Previously it has been extensively used in predicting two-phase fluid flows [25–27]. The VOF formulation relies on the fact that two or more fluids do not interpenetrate each other. For each phase considered in the model, a variable is introduced as the volume fraction of the phase in the computational cell. In each of the control volumes, the volume fractions of all phases sum up to unity. Because the temperature changes only slightly, all the fluid properties, except density, are considered as constants. The governing

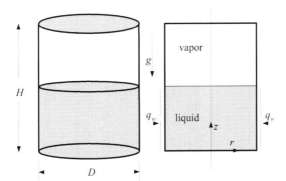

Figure 9.1 A partially filled tank.

equations from Chapter 5 are valid here as well, but are not repeated here. Only the level-set method for tracking the interface is briefly described.

The level-set method is a popular interface-tracking method for computing two-phase flows with topologically complex interfaces. This has similarities with the interface-tracking method of the VOF model. In the level-set method [28] the interface is captured and tracked by the level-set function φ, which is defined as an assigned distance from the interface. Accordingly, the interface is the zero level-set $\varphi(x, t)$ and can be expressed as $\Gamma = \{x | \varphi(x, t) = 0\}$ in a two-phase flow system:

$$\varphi(x, t) = \begin{cases} +|d| & \text{if } x \in \text{the primary phase} \\ 0 & \text{if } x \in \Gamma \\ -|d| & \text{if } x \in \text{the secondary phase} \end{cases} \tag{9.1}$$

where d is the distance from the interface. The normal and curvature of the interface, which are needed in the computation of the surface tension force, can be estimated by

$$\overrightarrow{n} = \frac{\nabla \varphi}{|\nabla \varphi|}\bigg|_{\varphi=0}, \kappa = \nabla \cdot \frac{\nabla \varphi}{|\nabla \varphi|}\bigg|_{\varphi=0} \tag{9.2}$$

The evolution of the level-set function is given in a similar fashion as for the VOF model:

$$\frac{\partial \varphi}{\partial t} + \nabla \cdot \left(\overrightarrow{V} \varphi\right) = 0 \tag{9.3}$$

The momentum equation can be written as

$$\frac{\partial \left(\rho \overrightarrow{V}\right)}{\partial t} + \nabla \cdot \left(\rho \overrightarrow{V} \overrightarrow{V}\right) = -\nabla p - \rho \beta \overrightarrow{g}(T - T_0)$$
$$+ \nabla \cdot \left[\mu_{\text{eff}} \left(\nabla \overrightarrow{V} + \nabla \overrightarrow{V}^{\text{T}}\right)\right] - \sigma \kappa \delta(\varphi) \nabla \varphi \tag{9.4}$$

In Eq. (9.4), $\delta(\varphi) = (1 + \cos(\pi \varphi / a))/2a$ if $|\varphi| < a$ and $a = 1.5h$ (where h is the grid spacing), otherwise $\delta(\varphi) = 0$. σ is the surface tension.

The boundary conditions are as follows, see also Fig. 9.1. The top and bottom surfaces of the tank are assumed to be flat and perfectly insulated. Heat-in-leak takes place at the cylindrical wet walls and an in-leak heat flux is assigned. The heating of LH$_2$ at the walls induces free convection currents (depending on the gravity value), with the warmer LH$_2$ in the wall region being forced to the upper regions of the liquid column.

Quiescent saturation conditions were assumed before imposing the heat flux at the cylindrical walls. The initial pressure is set to 1 atm, and the initial temperature corresponds to the saturation temperature (20.8K) at that pressure. The initial temperature is assumed uniform throughout the liquid and vapor. The pressure in the liquid is taken as a function of the height and density. No slip boundary conditions are imposed on the sidewalls. The top and bottom surfaces are assumed to be adiabatic. Because of the geometry, boundary conditions, and the physics of the problem, the flow is treated as axisymmetric.

The Rayleigh number, $Ra^* = g\beta\rho^2 c_p q_w L^4 / \mu\lambda^2$, which characterizes the flow type in the liquid, is 1.05×10^{12} for a gravity acceleration of $10^{-3} g_0$, where g_0 is the normal gravity (9.81 m/s^2). The magnitude of the Rayleigh number indicates that turbulent natural convection occurs in the liquid [29]. In zero-gravity conditions, laminar flow prevails and the Rayleigh number is zero, i.e., no buoyancy force acts and the surface tension becomes more important; the heat flux is 250 W/m^2. This value corresponds to the overall heat-in-leak in large insulated LH$_2$ tanks [30]. The fill level is 50% and the gravity values considered are microgravity of $10^{-3} g_0$ and zero gravity. The contact angle is assumed as 5 and 10 degrees, respectively, and the surface tension has the values 0.002 and 0.005 N/m, respectively.

The numerical implementation is very similar to what was described in Chapter 5 in terms of the number of grid points, choice of turbulence model, and control of the Courant number.

Validation of the formulation and computational methodology was carried out by considering the vaporization at the interface and a comparison with the experimental results in Ref. [31] for a liquid nitrogen tank. In the validation a cylindrical tank with a diameter of 0.201 m and a height of 0.213 m was used. It was filled up to 50% with liquid nitrogen, with a heat-in-leak of 1.2 W. Only nitrogen gas was considered to be present in the ullage space.

9.6.1 Results

The influence of gravity on the self-pressurization process was studied for a partially filled tank, in which the interface between the liquid and vapor is not flat under microgravity conditions [32,33]. Therefore, the steady-state distribution of the phases in the tank must be determined before the heat flux load is imposed. The fluid is then in equilibrium and treated as a viscous Newtonian fluid with constant properties (including density) and described by the regular time-dependent mass conservation and Navier–Stokes equations.

The distribution of the fluid in the tank at a microgravity value of $g = 9.81 \times 10^{-3}$ m/s$^2 = 10^{-3}$ g_0 (where g_0 is the normal gravity acceleration) and at zero gravity was calculated, as shown in Fig. 9.2. The contact angle was set as 5 degrees, as the attempt was to obtain the interface shape before applying heat flux. The liquid reaches the top wall along the tank wall because of the surface tension under zero gravity. The interface changes shape until it reaches a steady state from the initial flat interface. In Fig. 9.2, after the liquid has reached the top wall, it turns and moves to the center along the top wall. Liquid is accumulated at the center of the top wall like a droplet. The vapor fraction is symmetric but not completely neutral in the center of the tank. No parts of the walls remain dry. A thin liquid film appears on the top wall. In microgravity the surface tension is coupled with the remaining gravity and the coupled effect makes the interface not to be flat but curved.

The heat flux of 250 W/m^2 is then applied on the sidewall of the tank after the steady-state configuration has been achieved. The pressurization for different gravity levels is demonstrated in Fig. 9.3.

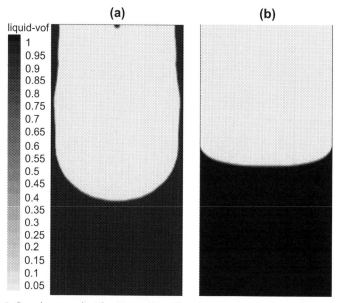

Figure 9.2 Steady-state distributions of liquid and vapor in a tank at different values of gravity: (a) 0 g_0 (zero gravity) and (b) 10^{-3} g_0 (microgravity). *vof*, volume fraction.

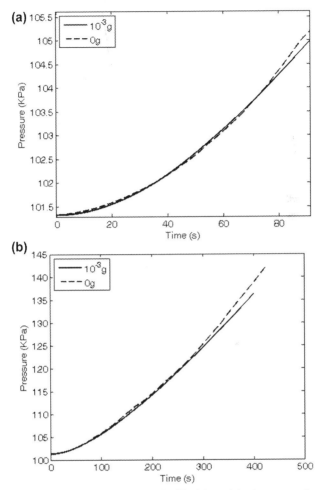

Figure 9.3 Self-pressurization in a partially filled liquid hydrogen tank at different gravities and periods: (a) initial period and (b) relatively long period.

The pressure rises more rapidly as the gravity level increases during the initial period, as shown in Fig. 9.3a. This is because under microgravity the area of the liquid–wetted wall is larger and it takes more time for the heat to be transferred to the interface without intensive convection. As time progresses (Fig. 9.3b), the pressure becomes larger and rises faster in zero gravity than that in microgravity. This is because superheated liquid starts to evaporate and small vapor regions emerge, grow, and detach from the tank wall. Fig. 9.4 shows the temperature and streamlines under zero gravity and a reduced gravity of $10^{-3}\ g_0$. Rounded isotherms are formed close to the

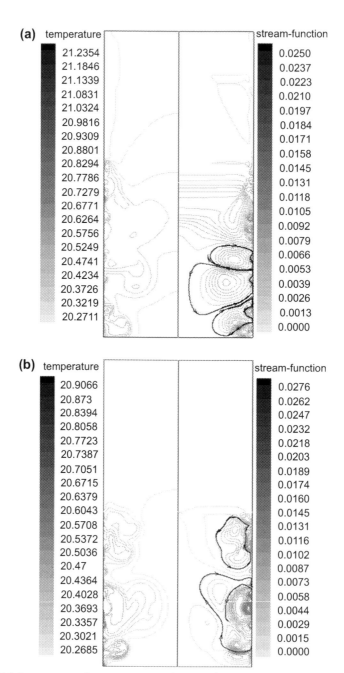

Figure 9.4 Isocontours of temperature and stream function during self-pressurization for two gravity conditions: (a) zero gravity (0 g_0) and (b) microgravity (10^{-3} g_0).

wall, whereas thermal stratification is observed slightly in the center part along the axis. Many separate flow loops are developed near the wall at different heights.

This phenomenon can be explained by considering Fig. 9.5.

The right side of each image is the zoom up of the black box on the left. The heat accumulated at the wall makes the temperature of the liquid higher and higher and evaporation begins to occur as the liquid becomes superheated. Small vapor regions start to form on the wall, which can be seen in Fig. 9.5, and the volume of the liquid is not equal to unity at the place where the rounded isotherms occur.

As the sidewall heating continues, the vapor region grows and detaches from the tank wall and then moves to the interface under the coupling of the buoyancy force and surface tension or forms a multiphase liquid–vapor foam. This can be seen in Figs. 9.6a and 9.7a. Fluid vortices are observed close to the vapor regions in Figs. 9.6b and 9.7b, especially at the location behind the motion of the vapor regions. Usually a clockwise rotation occurs near the tank wall just after the vapor leaves. The vortices force the vapor to perform a complex motion instead of moving upward directly. Because of the axial symmetry of the tank and the vortices, the vapor regions from the

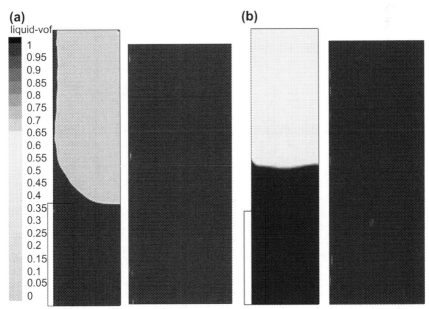

Figure 9.5 Volume fraction (vof) of the liquid at two different gravities: (a) 0 g_0 (zero gravity) and (b) 10^{-3} g_0 (microgravity).

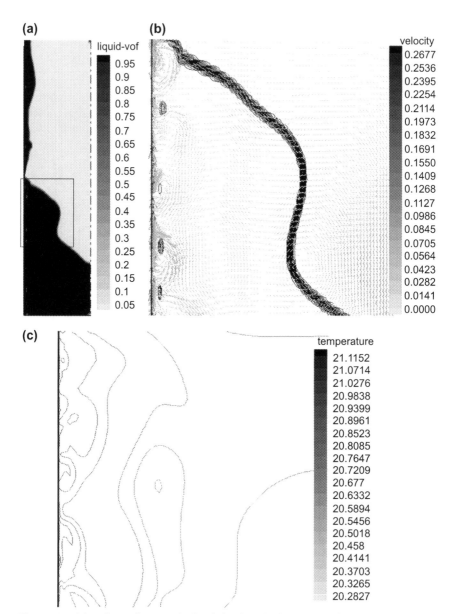

Figure 9.6 (a) Volume fraction (vof) of the liquid, (b) vector of velocity (m/s), and (c) temperature (K) of self-pressurization for 0 g_0 at 170 s.

tank wall merge at the midaxis at different heights with larger vapor regions and move up to the interface. Parts of the vapor arrive at the interface before merging during the flow motion. The velocity around the vortices is larger than the bulk fluid velocity. Isotherms are found around the vapor

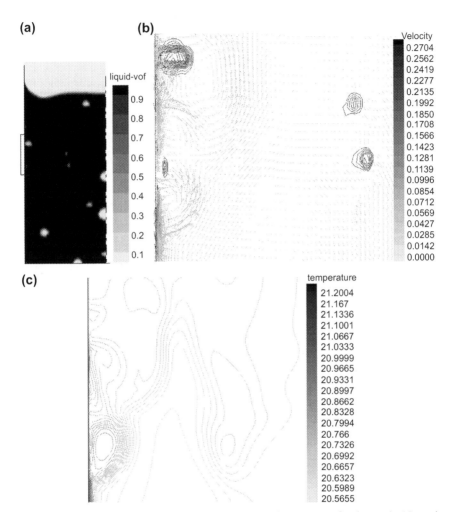

Figure 9.7 (a) Volume fraction (vof) of the liquid, (b) vectors of velocity (m/s), and (c) temperature (K) during self-pressurization for 10^{-3} g_0 at 170 s.

regions in Figs. 9.6c and 9.7c. As the heat cannot be easily carried up to the interface for evaporation, it accumulates at the tank wall. This heat then increases the temperature of the tank wall and the liquid propellant. The heat transfer efficiency is low as compared to natural convection. As time elapses, the enthalpy of the heated liquid exceeds the enthalpy according to the tank pressure, and accordingly, phase change starts to appear at the tank wall and evaporation becomes the main mode of heat transfer. The heat transfer coefficient increases and more heat is transferred into the tank. Furthermore, if the evaporated liquid adheres to the tank wall, hot spot forms may be produced. These will be dangerous for cryogenic liquid storage.

By comparing Figs. 9.6 and 9.7, it can be seen that the maximum velocity decreases as the gravity acceleration decreases, i.e., 0.2704 m/s for 10^{-3} g_0 and 0.2677 m/s for 0 g_0 at the same instant of time. This is because in microgravity the buoyancy force is still important and cannot be ignored. A higher velocity enhances the flow motion and the mixing results in a lower degree of thermal stratification. Therefore, the difference between the maximum and minimum temperatures is larger in zero gravity than that in microgravity.

The effects of contact angle and surface tension on self-pressurization will be discussed next. A microgravity of 10^{-3} g_0 with the same heat flux condition on the sidewall of the tank is considered. The previously shown results are valid for a contact angle of 5 degrees and a surface tension of 0.002 N/m. Fig. 9.8 shows the influence of the contact angle, whereas the influence of the surface tension is illustrated in Fig. 9.9. The pressure and its rise rate decrease as both the contact angle and surface tension increase. This is because more heat is used for phase change instead of heating the gas at the interface.

To explain this phenomenon in more detail, an analysis is presented. The heat flux partitioning model of Kurul and Podowski [34] has the following structure:

$$q_{wall} = q_c + q_q + q_e \qquad (9.5)$$

The first part is the single-phase heat transfer (convective heat flux):

$$q_c = A_1 h_{wallfcn}(T_{wall} - T_1) \qquad (9.6)$$

A_1 is the fraction of the wall surface influenced by the liquid. T_1 is the liquid temperature at the center of the wall adjacent to the computational cell and $h_{wallfcn}$ is the wall heat transfer coefficient calculated from the temperature wall function.

The quenching part q_q of the wall heat flux q_{wall} is transported by transient conduction during the period between the vapor bubble departure and the next vapor bubble formation at the same nucleation site. It is calculated as

$$q_q = A_2 h_{quench}(T_{wall} - T_1) \qquad (9.7)$$

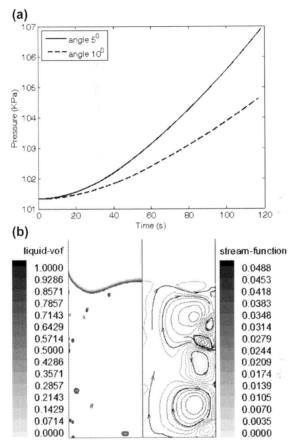

(a)

(b)

liquid-vof

stream-function

Figure 9.8 (a) Self-pressurization in a partially filled liquid hydrogen tank for different contact angles. (b) Volume fraction (vof) of the liquid and stream function during self-pressurization for a contact angle of 10 degrees at 170 s.

where A_2 is the fraction of the wall influenced by vapor bubbles formed at the wall and h_{quench} is the quenching heat transfer coefficient with the form

$$h_{quench} = 2f\sqrt{\frac{t_w \rho_l c_{pl} k_l}{\pi}} \tag{9.8}$$

The vapor bubble departure frequency f is determined from the following equation:

$$f = \sqrt{\frac{4g(\rho_l - \rho_v)}{3d_{bw}\rho_l}} \tag{9.9}$$

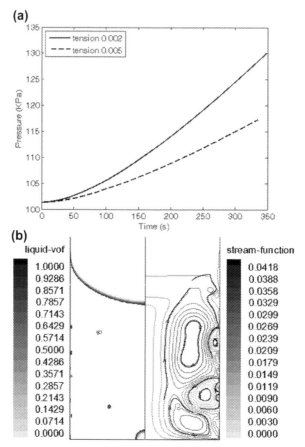

Figure 9.9 (a) Self-pressurization in a partially filled liquid hydrogen tank for different values of the surface tension. (b) Volume fraction (vof) of the liquid and stream function during self-pressurization for the surface tension of 0.005 N/m at 170 s.

The waiting time τ_w is the difference between the vapor bubble period and the growth time τ_g,

$$t_w = 1/f - t_g \tag{9.10}$$

The heat flux of vaporization q_e can be computed by the following equation:

$$q_e = \frac{\pi}{6}d_{bw}^3\rho_v nfL_H \tag{9.11}$$

where d_{bw}, n, and f are the departure diameter of a vapor bubble, the number of active nucleation sites, and the vapor bubble departure frequency, respectively. The departure diameter is calculated by the equation studied by Kirichenko et al. [35]:

$$d_{bw} = C_{bw} \left(\frac{3}{2} \frac{\sigma}{g(\rho_l - \rho_v)} \frac{2\sigma T_{sat}}{L_H \rho_v \Delta T_{sup}} \right)^{1/3} \qquad (9.12)$$

The active nucleation site density n is correlated to the wall superheat as

$$n = C_n \left(\frac{L_H \rho_v \Delta T_{sup}}{\sigma T_{sat}} \right)^m \qquad (9.13)$$

In Eq. (9.13), C_n and m are determined by

$$\begin{cases} P/P_{cr} > 0.04 \sim 0.06: C_n = 10^{-7}, m = 2 \\ P/P_{cr} < 0.02 \sim 0.04: C_n = 6.25 \times 10^{-14}, m = 3 \end{cases} \qquad (9.14)$$

The evaporation part q_e contributes to the pressure rise. The other two parts of the wall heat flux mainly increase the temperature of the whole system. From the above-mentioned description, the contact angle and surface tension affect the evaporation heat flux through the vapor bubble dynamics. Zeng et al. [36] introduced a lattice Boltzmann method to simulate the bubble growth in pool boiling and concluded that the influence of the contact angle on the departure diameter of a vapor bubble can be neglected. However, the departure frequency decreases as the contact angle increases. Thus the heat flux of vaporization decreases according to Eq. (9.11) and this results in a decrease in the pressure magnitude and its rise rate.

By combining Eqs. (9.9)−(9.14), the relation between the evaporation heat flux and the surface tension is obtained as

$$q_e \propto \sigma^{2-(m+\frac{1}{3})} \qquad (9.15)$$

If m = 2 or 3, the exponent is negative. Therefore, the heat flux q_e decreases as the surface tension σ increases, as shown in Fig. 9.9. Less heat is transferred to the vapor part through vaporization, which contributes to the pressure rise at the same wall heat flux.

Vortices are also observed at the location just behind the vapor bubble. The direction of the loops is reversed compared to the bulk fluid circulation. The warm liquid heated by the wall moves no longer just up along the wall to the interface for evaporation and falls down again in the middle in the presence of bubbles. The motion is much more complicated.

REFERENCES

[1] Dhindaw BK. Solidification under microgravity. Sadhana 2001;26(1 & 2):59−69.

[2] Ozawa S, Yoshizawa J, Saito T, Motegi T. Microstructure of Nd-Fe-B alloys solidified under microgravity conditions. Mater Trans JIM 2000;41(9):1121−4.

[3] Mirihanage WU, Browne DJ, Sturz L, Zimmermann G. Numerical modelling of the material science lab − low gradient furnace (MSL-LGF) microgravity directional solidification experiments on the columnar to equiaxed transition. IOP Conf. Series: Materials Science and Engineering, 27; 2011. paper no. 012010.

[4] Murphy AG, Li J, Janson O, Verga A, Browne DJ. Microgravity and hypergravity observations of equiaxed solidification of Al-Cu alloys using in-situ X-radiography recorded in real-time on board a parabolic flight. Materials Science Forum 2014;70−791:52−8.

[5] Giangi M, Stella F, Leonardi E, De Vahl Davis G. A numerical study of solidification in the presence of a free surface under microgravity conditions. Num. Heat Transfer Part A 2002;41:579−95.

[6] Hofmann DC, Roberts SN. Microgravity metal processing: from undercooled liquid to bulk metallic glasses. Microgravity 2015;1. paper no. 15003.

[7] An CY, Toledo RC, Boschetti C, Riberio MF, Bandeira IN. Solidification of lead tin and lead telluride eutectic alloys in microgravity. Microgravity Sci Technol 2014;25(5):267−73.

[8] Wang J, Rao T, Li X, Pei Y. Analysis of influence of gravity on convection heat transfer in manned spacecraft during terrestrial test. Int J Mech Aerosp Ind Mechatron Manuf Eng 2012;6(9):1908−13.

[9] Kostoglou M, Evgendis SP, Zacharias KA, Karapantsios TD. Heat transfer from small objects in microgravity: experiments and analysis. Int J Hat Mass Transf 2011;54:3323−33.

[10] Chen Y, Sohan CB, Peterson GP. Review of condensation heat transfer in microgravity environments. AIAA J Thermophys Heat Transf 2006;20(3):353.

[11] Chow LC, Parish RC. Condensation heat transfer in a microgravity environment. AIAA J Thermophys Heat Transf 1987;2(1):82.

[12] Liu Y, Wang W. Condensation heat transfer inside a tube in a microgravity environment. J Therm Sci 1996;5(3):184−9.

[13] Mishkinis D, Ochterbeck JM. Analysis of tubeside condensation in microgravity and earth-normal gravity. In: Proc. Vth Minsk International Seminar heat pipes, heat pumps refrigerators; 2003. p. 36−53.

[14] Lee H, Mudawar I, Hasan MM. Experimental and theoretical investigation of annular flow condensation in microgravity. Int J Heat Mass Transf 2013;61:293−309.

[15] Ghosh P, Sarkar A, Das S. Analysis of film condensation along a vertical flat plate under sinusoidal G-Jitter. Microgravity Sci Technol 2013;25:95−102.

[16] Sunden B. Introduction to heat transfer. Southampton, UK: WIT Press; 2012.

[17] Wang L, Zhu K, Xie F, Ma Y, Li Y. Prediction of pool boiling heat transfer for hydrogen in microgravity. Int J Heat Mass Transf 2016;94:465−73.

[18] Yang Y, Pan LM, Xu JJ. Effects of microgravity on Marangoni convection and growth characteristic of a single bubble. Acta Astronaut 2014;100:129−39.

[19] Baltis C, Celata GP, Cumo M, Saraceno L, Zummo G. Gravity influence on heat transfer rate in flow boiling. Microgravity Sci Technol 2012;24:2013−213.

[20] Fu J, Sunden B, Chen X, Huang Y. Influence of phase change on self-pressurization in cryogenic tanks under microgravity. Appl Therm Eng 2015;87:225−33.

[21] Fu J, Sunden B, Chen XQ. Analysis of self-pressurization phenomenon in a cryogenic fluid storage tank with VOF method. In: Proceedings of the ASME 2013 International Mechanical Engineering Congress & Exposition; 2013. IMECE2013−63209.

[22] Fu J, Sunden B, Chen XQ. Influence of wall ribs on the thermal stratification and self-pressurization in a cryogenic liquid tank. Appl Therm Eng 2014;73:1421–31.

[23] Albadawi A, Donoghue DB, Robinson AJ, Murray DB, Delauré YMC. Influence of surface tension implementation in volume of fluid and coupled volume of fluid with level set methods for bubble growth and detachment. Int J Multiph Flow 2013;53:11–28.

[24] Hirt CW, Nichols BD. Volume of fluid (VOF) method for the dynamics of free boundaries. J Comput Phys 1981;39:201–25.

[25] Luckmann AJ, Alves MVC, Barbosa Jr JR. Analysis of oil pumping in a reciprocating compressor. Appl Therm Eng 2009;29:3118–23.

[26] Li G, Frankel S, Braun JE, Groll EA. Application of CFD model to two-phase flow in refrigeration distributors. J HVAC R Res 2005;11:45–62.

[27] Sussman M, Smith KM, Hussaini MY, Ohta M, Zhi-Wei R. J Comput Phys 2007;221:469–505.

[28] Osher S, Sethian JA. Fronts propagating with curvature-dependent speed: algorithms based on Hamilton-Jacobi formulations. J Comput Phys 1988;79:12–49.

[29] Gursu S, Sherif SA, Veziroglu TN, Sheffield JW. Analysis and optimization of thermal stratification and self-pressurization effects in liquid hydrogen storage systems, Part 1: model development. ASME J Energy Resour Technol 1993;115:221–7.

[30] Kumar SP, Prasad BVSSS, Venkatarathnam G, Ramamurthi K, Murthy SS. Influence of surface evaporation on stratification in liquid hydrogen tanks of different aspect ratios. Int J Hydrogen Energy 2007;32:1954–60.

[31] Seo M, Jeong S. Analysis of self-pressurization phenomenon of cryogenic fluid storage tank with thermal diffusion model. J Cryog 2010;50:549–55.

[32] Stark JA, Bradshow RD, Blatt MH. Low-gravity fluid behavior technology summaries. 1974. NASA-CR-134748, USA.

[33] Ostrach S. Low-gravity fluid flows. Ann Rev Fluid Mech 1974;14:313–45.

[34] Kurul N, Podowski MZ. Multidimensional effects in forced convection subcooled boiling. In: Proc. 9th Int. Heat Transfer Conference, vol. 1; 1990. p. 21–6.

[35] Kirichenko YA, A Slobozhanin L, Shcherbakova NS. Analysis of quasi-static conditions of boiling onset and bubble departure. J Cryog 1983;23:110–2.

[36] Zeng JB, Li LJ, Liao Q, Jiang FM. Simulation of bubble growth process in pool boiling using lattice Boltzmann method. Acta Phys Sin 2011;60:066401.

CHAPTER 10

Computational Methods for the Investigations of Heat Transfer Phenomena in Aerospace Applications

10.1 INTRODUCTION

Computational methods started to have a significant impact on the analysis of aerodynamics and its design in the late 1960s. The so-called panel methods were introduced, and these were based on the distribution of surface singularities on a given configuration. Potential flows (nonviscous flows) around bodies could be solved by these methods. Additional capabilities were added later to the surface panel methods and then it was possible to include higher order more accurate formulations, lifting capability, unsteady flows, and coupling with various boundary layer formulations. However, the panel methods could not offer accurate solutions for high-speed nonlinear flows of current interest, and thus more sophisticated models of the flow field equations had to be developed. Gradually, this development has led to what is now called computational fluid dynamics (CFD). For further details, see Ref. [1].

CFD is an interdisciplinary branch of science and engineering with a broad spectrum of applications. Fluid flow, heat transfer, mass transfer, combustion, and chemical reactions appear in most aspects of modern life and are of significance in automotive, space and aviation, and chemical and process industries, as well as in atmospheric science, energy, medicine, microtechnology, and nanotechnology. The development and applications of CFD have been tremendous, and CFD is used both as a modeling tool and in R&D in many industries nowadays. Besides it continues to be developed for new challenges and is being used in basic research at universities.

In aerospace applications, the fluid flow is compressible and the fluid density varies with its pressure. The flow speed is commonly high and the

Heat Transfer in Aerospace Applications
ISBN 978-0-12-809760-1
http://dx.doi.org/10.1016/B978-0-12-809760-1.00010-7

Mach number is greater than 0.3. Subsonic compressible flows have been found to have a Mach number between 0.3 and 0.8. The relationship between pressure and density is weak, and no shocks will be computed within the flow. Highly compressible flows have a Mach number greater than 0.8. The pressure strongly affects the density, and shocks are possible. Compressible flows can be either transonic (0.8 < Ma < 1.2) or supersonic (1.2 < Ma < 3.0). In supersonic flows, pressure effects are transported only downstream. The upstream flow is not affected by conditions and obstructions downstream.

The total temperature, T_t, is a key parameter and is the sum of the static temperature and the dynamic temperature. There are two ways to calculate the total temperature, see Eq. (10.1):

$$T_t = T + \frac{V_i^2}{2c_p} \quad \text{and} \quad T_t = T\left(1 + \frac{\gamma - 1}{2}Ma^2\right) \tag{10.1}$$

where V is the velocity and c_p is the gas specific heat.

The total pressure, P_t, is another useful quantity for running compressible analyses. It is the sum of the static pressure and the dynamic pressure.

In general, compressible flow analyses are much more sensitive to the applied boundary conditions and material properties than incompressible analyses. If the applied settings do not define a physically real flow situation, then the analysis can be very unstable and may fail to reach a converged solution. Proper specification of the boundary conditions and material properties will greatly improve the chances of a successful analysis.

To include heat transfer in a compressible analysis, it is recommended to apply the total (stagnation) temperature boundary conditions instead of static temperatures at the inlets. Total temperature should also be applied to any solids or walls with known temperature conditions. When there is heat transfer in a compressible analysis, viscous dissipation, pressure work, and kinetic energy terms are calculated. It is very important that the total temperature is specified correctly.

In this chapter a brief summary of the CFD methods, including turbulence modeling; associated problems; and limitations is provided. Examples of CFD applications in aerospace engineering are also provided. Commercially available computer codes and in-house codes are briefly described.

10.2 GOVERNING EQUATIONS

All the governing differential equations of mass conservation, transport of momentum, energy, and mass fraction of species can be cast into a general partial differential equation as [2,3],

$$\frac{\partial \rho \phi}{\partial t} + \frac{\partial}{\partial x_j} \rho \phi u_j = \frac{\partial}{\partial x_j}\left(\Gamma \frac{\partial \phi}{\partial x_j}\right) + S \qquad (10.2)$$

where ϕ is an arbitrary dependent variable, such as the velocity components and temperature; Γ, the generalized diffusion coefficient; and S, the source term for ϕ. The general differential equation consists of four terms. From left to right in Eq. (10.2), they are referred to as the unsteady term, the convection term, the diffusion term, and the source term.

10.3 NUMERICAL METHODS TO SOLVE THE GOVERNING DIFFERENTIAL EQUATIONS

Some numerical methods have been established to solve the governing equations of fluid flow and heat transfer problems. They are the finite difference method (FDM) [4], the finite volume method (FVM) [2,3], the finite element method (FEM) [5,6], the control volume finite element method (CVFEM) [7], and the boundary element method (BEM) [8]. In this chapter, only some details of the FVM will be presented.

10.3.1 The Finite Volume Method

In the FVM the domain is subdivided into a number of so-called control volumes. The integral form of the conservation equations is applied to each control volume. At the center of the control volume a node point is placed. At this node the variables are located. The values of the variables at the faces of the control volumes are determined by interpolation. The surface and volume integrals are evaluated by quadrature formulas. Algebraic equations are obtained for each control volume. In these equations, values of the variables for neighboring control volumes appear.

The FVM is very suitable for complex geometries, and the method is conservative as long as surface integrals are the same for control volumes sharing boundary.

The FVM is a popular method particularly for convective flow and heat transfer. It is also applied in several commercial CFD codes. Further details

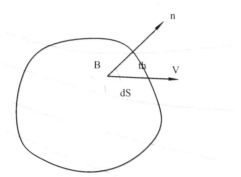

Figure 10.1 A control volume.

can be found in Ref. [2,3]. A brief illustration is presented in the following section, and an arbitrary control volume is shown in Fig. 10.1.

A formal integration of the general equation across the control volume reads

$$\iiint_V \frac{\partial \rho U_j \phi}{\partial x_j} dV = \iiint_V \frac{\partial}{\partial x_j}\left(\Gamma_\phi \frac{\partial \phi}{\partial x_j}\right) dV + \iiint_V S_\phi dV \qquad (10.3)$$

Then by applying the Gaussian theorem or the divergence theorem,

$$\iint_S \rho \phi \vec{U} \cdot d\vec{S} = \iint_S \Gamma_\phi \nabla \phi \cdot d\vec{S} + \iiint_V S_\phi dV \qquad (10.4)$$

By summing up all the faces of the control volume, the equation becomes

$$\sum_{f=1}^{nf} \phi_f C_f = \sum_{f=1}^{nf} D_f + S_\phi \Delta V \qquad (10.5)$$

where the convection flux, C_f; the diffusion flux, D_f; and the scalar value of the arbitrary variable ϕ at a face, Φ_f have to be determined.

10.3.1.1 Convection-Diffusion Schemes

To achieve physically realistic results and stable iterative solutions, a convection–diffusion scheme needs to possess the properties of conservativeness, boundedness, and transportiveness. The upstream, hybrid, and power-law discretization schemes all possess these properties and are generally found to be stable but they suffer from numerical or false diffusion in multidimensional flows if the velocity vector is not parallel with one of the coordinate directions. The central difference scheme lacks transportiveness and is known to give unrealistic solutions at large Peclet numbers. Higher order schemes such as QUICK (quadratic upstream interpolation for convective kinematics) and van Leer may minimize the numerical diffusion but are less numerically stable. Also implementation of boundary conditions can be somewhat problematic with higher order schemes and the computational demand can be extensive because additional grid points are needed, and the expressions for the coefficients in Eq. (10.5) become more complex.

10.3.1.2 Source Term

The source term S may in general depend on the variable ϕ. In the discretized equation, it is desirable to account for such a dependence. Commonly the source term is expressed as a linear function of ϕ.

At the grid point P, S is then written as

$$S = S_C + S_P\phi \tag{10.6}$$

In order to prevent divergence, it is required that S_P is negative. The linearization procedure in Eq. (10.6) is commonly used.

10.3.1.3 Solution of the Discretized Equations

The discretized equations have the form of Eq. (10.5) with the ϕ values at the grid points as unknowns. For boundaries not having fixed ϕ values, the boundary values can be eliminated by using given or fixed conditions of the fluxes at such boundaries. Gauss elimination is a direct method to solve algebraic equations. For one-dimensional cases the coefficients form a tridiagonal matrix and an efficient algorithm called the Thomas algorithm or the tridiagonal matrix algorithm (TDMA) is achieved. For two-dimensional and three-dimensional (3D) cases, direct methods require large computer memory and computer time. Iterative methods are, therefore, used to solve the algebraic equations. A popular method is a line-by-line technique combined with a block correction procedure. The equations along the chosen line are solved by the TDMA. Iterative methods are also needed because the equations are nonlinear and sometimes interlinked.

In many situations, e.g., turbulent forced convection, the change in the value of ϕ from one iteration to another is so high that convergence in the iterative process is not achieved. To circumvent this issue and to reduce the magnitude of the changes, underrelaxation factors (between 0 and 1) are introduced.

10.3.1.4 The Pressure in the Momentum Equations

In the momentum equations, a pressure gradient term appears in each coordinate direction (i.e., a source term S). If these gradients are known, the discretized equations for the flow velocities would follow the same procedure as for any scalar. However, in general the pressure gradients are not known but have to be determined as part of the solution. Thus the pressure and velocity fields are coupled and the continuity equation (mass conservation equation) has to be used to develop a strategy.

There are also other related difficulties in solving the momentum and continuity equations. It has been shown that if the velocity components and the pressure are straightforwardly calculated at the same grid points, some physically unrealistic fields, such as checkerboard solutions, may arise in the numerical solution. A remedy to this problem is to use staggered grids. The velocity components are then given staggered or displaced locations. These locations are such that they lie on the control volume faces that are perpendicular to them. All other variables are calculated at the ordinary grid points. Another remedy is to use a nonstaggered or collocated grid where all variables are stored at the ordinary grid points. A special interpolation scheme is then applied to calculate the velocities at the control volume faces. Most commonly, the so-called Rhie—Chow interpolation method is applied [9].

10.3.1.5 Procedures for Solution of the Momentum Equations

As mentioned in the preceding section, the velocity and pressure fields are coupled. Thus a strategy has to be developed in the solution procedure of the momentum equations. The oldest algorithm is the SIMPLE (semi-implicit method for pressure-linked equations) algorithm. A pressure field is guessed and then the momentum equations are solved for this pressure field, resulting in a velocity field. Then a pressure correction and velocity corrections are introduced. From the continuity equation an algebraic equation for the pressure correction can be obtained. The velocity corrections are related to the pressure corrections and the coefficients linking the velocity corrections to the pressure corrections depend on the chosen algorithm.

The momentum equations are then solved again but now with the corrected pressure as the guessed pressure. New velocities are obtained and new pressure and velocity corrections are calculated. The whole process is repeated until convergence is obtained.

There are other similar algorithms available today; SIMPLEC (SIMPLE consistent) and SIMPLEX (SIMPLE extended) are common. They differ from SIMPLE mainly in the expression for coefficients linking velocity corrections to the pressure correction [10].

Another algorithm named PISO (pressure implicit splitting operators) [11] has become popular more recently. Originally, it is a pressure–velocity coupling strategy for unsteady compressible flow. Compared to SIMPLE, it involves one predictor step and two corrector steps.

Still another algorithm is SIMPLER (SIMPLE revised). Here the continuity equation is used to derive a discretized equation for the pressure. The pressure correction is then only used to update the velocities through the velocity corrections.

10.3.1.6 Convergence

The solution procedure is in general iterative and then some criterion must be used to decide when a converged solution has been reached. One method is to calculate residuals R as

$$R = \sum_{NB} a_{NB}\phi_{NB} + b - a_P\phi_P \qquad (10.7)$$

for all variables. NB means neighboring grid points, e.g., E, W, N, S (East, West, North, South). If the solution is converged, $R = 0$ everywhere. Practically, it is often stated that the largest value of the residuals [R] should be less than a certain number. If this is achieved the solution is said to be converged.

10.3.1.7 Number of Grid Points and Control Volumes

The widths of the control volumes do not need to be constant or the successive grid points do not have to be equally spaced. Often it is desirable to have a uniform grid spacing. Also it is required that a fine grid is employed where steep gradients appear, whereas a coarse grid spacing may suffice where slow variations occur. The various turbulence models require certain conditions on the grid structure close to solid walls. The so-called high- and low-Reynolds-number versions of these models demand different conditions.

In general, it is recommended that the solution procedure is carried out on several grids with different fineness and varying degrees of nonuniformity. Then it might be possible to estimate the accuracy of the numerical solution procedure.

Adaptive grid techniques might be beneficial to increase the resolution in vital areas, such as resolving the pressure jump after a shockwave. In modeling hypersonic flows, special care must be taken in meshing external flows, especially near-shock conditions, as a fine mesh is required to capture shock effects.

10.3.1.8 Complex Geometries

CFD methods based on the Cartesian, cylindrical, or spherical coordinate systems have limitations in complex or irregular geometries. Using the Cartesian, cylindrical, and/or spherical coordinates means that the boundary surfaces are treated in a stepwise manner. To overcome this problem, methods based on body-fitted or curvilinear orthogonal and nonorthogonal grid systems are needed. Such grid systems may be unstructured, structured, or block structured or composite. Because the grid lines follow the boundaries, boundary conditions can more easily be implemented.

There are also some disadvantages with nonorthogonal grids. The transformed equations contain more terms and the grid nonorthogonality may cause unphysical solutions. Vectors and tensors maybe defined as Cartesian, covariant, contravariant, and physical or nonphysical coordinate oriented. The arrangement of the variables on the grid affects the efficiency and accuracy of the solution algorithm.

Grid generation is an important issue, and today, most commercial CFD packages have their own grid generators, and several grid generation packages, compatible with some CFD codes, are also available. The interaction with various CAD (computer-aided design) packages is also an important issue today.

Further information on treating complex geometries can be found in Refs. [12,13].

10.4 THE CFD APPROACH

The FVM described earlier is a popular method particularly for convective flow and heat transfer. It is also applied in several commercial CFD codes. In heat transfer equipment such as heat exchangers, both the laminar and turbulent flows are of interest. Although the laminar convective flow and heat transfer can be simulated, the turbulent flow and heat transfer normally require modeling approaches in addition. By turbulence modeling, the goal is to account for all the relevant physics using as simple a mathematic model as possible. This section gives a brief introduction to the modeling of turbulent flows.

The instantaneous mass conservation, momentum, and energy equations form a closed set of five unknowns u, v, w, p, and T. However, the computing requirements in terms of resolution in space and time for directly solving the time-dependent equations of fully turbulent flows at high Reynolds numbers [the so-called direct numerical simulation (DNS) calculations] are enormous and major developments in computer hardware are needed. Thus DNS is more viewed as a research tool for relatively simple flows at moderate Reynolds number and supercomputer calculations are required. Meanwhile, practicing thermal engineers need computational procedures to provide information on the turbulent processes, but avoiding the need to predict effects of every eddy in the flow. This calls for information about the time-averaged properties of the flow and temperature fields (e.g., mean velocities, mean stresses, mean temperature). Usually, a time-averaging operation, called Reynolds decomposition, is carried out. Every variable is then written as the sum of a time-averaged value and a superimposed fluctuating value. In the governing equations, additional unknowns appear: six for the momentum equations and three for the temperature field equation. The additional terms in the differential equations are called turbulent stresses and turbulent heat fluxes, respectively. The task of turbulence modeling is to provide procedures to predict the additional unknowns, i.e., the turbulent stresses and turbulent heat fluxes, with sufficient generality and accuracy. Methods based on the Reynolds-averaged equations are commonly referred to as the RANS (Reynolds-averaged Navier–Stokes equations) methods. Large eddy simulation (LES) lies between the DNS and RANS approaches in terms of computational demand. Like DNS, 3D simulations are carried out over many time steps but only the larger eddies are resolved. An LES grid can be coarser in space and the time steps can be larger than that for DNS, as the small-scale fluid motions are treated by the so-called subgrid-scale (SGS) model.

10.4.1 Turbulence Models

The most common turbulence models for industrial and aerospace applications are classified as

- zero-equation models
- one-equation models
- two-equation models
- Reynolds stress models
- algebraic stress models (ASMs)
- LESs

The first three models in this list account for the turbulent stresses and heat fluxes by introducing a turbulent viscosity (eddy viscosity) and a turbulent diffusivity (eddy diffusivity). Linear and nonlinear models exist [14–16]. The eddy viscosity is usually obtained from certain parameters representing the fluctuating motion. A popular one-equation model is the

Spalart—Allmaras model [17] in which a transport equation is solved for the eddy viscosity. It is mostly used for aerospace and turbomachinery applications but not very common for heat exchangers. In two-equation models, these parameters are determined by solving two additional differential equations. However, one should remember that these equations are not exact but approximate and involve several adjustable constants. Models using the eddy viscosity and eddy diffusivity approach are isotropic in nature and cannot evaluate nonisotropic effects. Various modifications and alternate modeling concepts have been proposed. Examples of models of this category are the k-ε and k-ω models in high- or low-Reynolds-number versions as well as in linear and nonlinear versions. A lately popular model is the so-called V2F model introduced by Durbin [18]. It extends the use of the k-ε model by incorporating near-wall turbulence anisotropy and nonlocal pressure-strain effects, while retaining a linear eddy viscosity assumption. Two additional transport equations are solved: one for the velocity fluctuation normal to walls and another for a global relaxation factor. More recently the shear stress transport (SST) k-ω model by Menter [19] has become popular, as it uses a blending function of gradual transition from the standard k-ω model near solid surfaces to a high-Reynolds-number version of the k-ε model far away from solid surfaces. Accordingly, it accurately predicts the onset and the size of separation under adverse pressure gradients.

In the Reynolds stress equation models (RSMs), differential equations for the turbulent stresses (the Reynolds stresses) are solved and directional effects are naturally accounted for. Six modeled equations (i.e., not exact equations) for the turbulent stress transport are solved together with a model equation for the turbulent scalar dissipation rate ε. The RSM models are quite complex and require large computing efforts and are therefore not widely used for industrial flow and heat transfer applications, such as in heat exchangers.

The ASM and explicit ASM (EASM) present an economic way to account for the anisotropy of the turbulent stresses without solving the Reynolds stress transport equations. An idea is that the convective and diffusive terms are modeled or even neglected and then the Reynolds stress equations are reduced to a set of algebraic equations.

For calculation of the turbulent heat fluxes, most commonly, a simple eddy diffusivity (SED) concept is applied. The turbulent diffusivity for heat transport is then obtained by dividing the turbulent viscosity by a turbulent Prandtl number. Such a model cannot account for nonisotropic effects in the thermal field but still this model is frequently used in engineering applications. There are some models presented in the literature to account for nonisotropic heat transport, e.g., the generalized gradient diffusion hypothesis (GGDH) and the WET (wealth = earnings × time) method. These higher order models require that the Reynolds stresses are calculated accurately by taking nonisotropic effects into account. If not, the performance may not be improved. In addition, partial differential equations can be formulated for the three turbulent heat fluxes but numerical solutions of these modeled equations are rarely found. Further details can be found in, e.g., Ref. [20].

In the LES model the time-dependent flow equations are solved for the mean flow and the largest eddies, while the effects of the smaller eddies are modeled. The LES model has been expected to emerge as the future model for industrial applications but it is still limited to relatively low Reynolds number and simple geometries. Handling wall-bounded flows with focus on the near-wall phenomena such as heat and mass transfer and shear at high Reynolds number presents a problem because of the near-wall resolution requirements. Complex wall topologies also present problems for LES.

Nowadays, models obtained by combining the LES and RANS-based methods have been suggested. Such models are called hybrid models and the detached eddy simulation (DES) [21] is an example.

In an article [22], it was stated that the prospects are good for steady problems with RANS turbulence modeling to be solved accurately even for very complex geometries if technologies for solution adaptation become mature for large 3D problems. It was also conjectured that the future rate of growth for supercomputers will only be half of the rate in the past 20 years. This is expected to slow down the pure LES reliability for aerospace applications at high Reynolds numbers but the hybrid RANS–LES approaches were judged to have a great potential. Also a breakthrough in turbulence modeling to predict separation and laminar–turbulent transition was not foreseen to appear in the near future.

10.4.2 Wall Effects

There are two standard procedures to account for wall effects in numerical calculations of turbulent flow and heat transfer: one is to employ low-Reynolds-number modeling procedures and the other is to apply the wall function method. The wall functions approach includes empirical formulas and functions linking the dependent variables at the near-wall cells to the corresponding parameters on the wall. The functions are composed of laws of the wall for the mean velocity and temperature, and formulas for the near-wall turbulent quantities. The accuracy of the wall function approach is increasing with increasing Reynolds number. In general the wall function approach is efficient and requires less CPU time and memory size but it becomes inaccurate at low Reynolds numbers. When low-Reynolds-number effects are important in the flow domain, the wall function approach ceases to be valid. The so-called low-Reynolds-number versions of the turbulence models are introduced and the molecular viscosity appears in the diffusion terms. In addition, damping functions are introduced. Also the so-called two-layer models have been suggested to enhance the wall treatment. The transport equation for the turbulent kinetic energy is solved, whereas an algebraic equation is used for, e.g., the turbulent dissipation rate.

10.4.3 CFD Codes

Several industries and engineering and consulting companies worldwide are nowadays using the commercially available general-purpose so-called CFD codes for simulation of fluid flow, heat and mass transfer, and combustion in aerospace applications. Among these codes are the ANSYS FLUENT, ANSYS CFX, CFD++, and STAR-CCM+. Also many universities and research institutes worldwide apply commercial codes, besides using their in-house developed codes. Nowadays, open-source codes such as OpenFOAM are also freely available. There are also specialized codes such as DPLR (data-parallel line relaxation) and VULCAN (viscous upwind algorithm for complex flow analysis) for hypersonic flows. DPLR is designed for supersonic and hypersonic flows under nonequilibrium conditions, whereas VULCAN is designed for internal flows in scramjet engines. Further details and brief descriptions of the codes can be found in Ref. [23].

However, to successfully apply such codes and to interpret the computed results, it is necessary to understand the fundamental concepts of computational methods. Important issues are also the description of complex geometries and the generation of suitable grids. The commercial codes commonly have their own grid generation tool, e.g., ANSYS ICEM, but stand-alone software such as Pointwise are also popular. The codes are generally also compatible with various CAD tools.

10.5 TOPICS NOT TREATED

There are several additional topics that are of importance in CFD modeling and simulations of aerospace problems. Among them, topics that are not being treated in this chapter are

- implementation of boundary conditions
- adaptive grid methods
- local grid refinements
- solution of algebraic equations
- convergence and accuracy
- parallel computing
- animation

10.6 EXAMPLES

10.6.1 Chemical Nonequilibrium Turbulent Flow in a Scramjet Nozzle

In this section, details and results from a numerical investigation of nonequilibrium flow in a scramjet single expansion ramp nozzle are presented with a chemical reaction model including seven species and eight finite rate steps [24]. The generic geometry of the scramjet nozzle is depicted in Fig. 10.2. The flow is internal at high temperature and pressure and external at high velocity and low pressure.

The internal fluid goes through a diverging nozzle before being mixed with the external flow. The top wall of the nozzle is designed for maximum trust according to the characteristic line method, and it is longer than the bottom wall making the nozzle to be a single expansion ramp nozzle. The total length of the nozzle is $L = 18.54\ H_1$, the length of the bottom wall is $L_s = 3.12\ H_1$, and the diffusive angle of the bottom wall $\theta = 6°$, where H_1 is the height of the nozzle inlet.

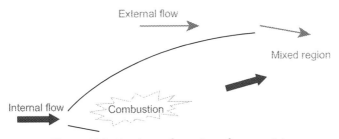

Figure 10.2 Nozzle configuration of a scramjet.

The FVM with a fully implicit scheme was used to solve the conservative unsteady RANS equations, including variations in density, pressure, velocities, mass fractions, and total and thermal energy per unit mass, with appropriate source terms. The heat fluxes are contributed by the species transport. Turbulence is handled by RNG (Re-Normalisation Group) k-ε model. The relation between the thermal energy per unit mass and pressure and density is handled as for nonequilibrium. A second-order upwind scheme is adopted in general but for the k-ε equations first-order upwind scheme was used. Nonequilibrium wall functions were employed to link the viscous region near the wall with the fully turbulent developed region. To consider the effect of the external flow on the nozzle performance, the domain of the external flow was extended. To describe the complicated flow field a multiblock technique was adopted for the structured grid generation, including local refinements. The grid is shown in Fig. 10.3.

The boundary conditions were for external flow the far field pressure was prescribed, no-slip and adiabatic walls were assumed, and total pressure, temperature, Ma number, and mass fractions were given at the nozzle inlet. Finite rate chemistry reaction models for hydrogen combustion in air were used. Seven species were considered, i.e., H_2O, OH, O_2, H, H_2, O, and N_2. Eight reactions were considered.

10.6.1.1 Some Results

Fig. 10.4 shows the total temperature contours of the chemical nonequilibrium flow inside the nozzle. Because of combustion the total temperature sharply increases at the inlet of the nozzle. Then the increasing trend is diminished.

A typical picture of the mass fraction distribution of OH is shown in Fig. 10.5. At first it increases and then it is reduced in the main flow direction. The variation is nonmonotonic because chemical reactions occur close to the wall.

Fig. 10.6 presents an example of the mass fraction distributions of H_2O, O_2, H_2, and N_2 along the nozzle walls. The distribution of H_2 and O_2

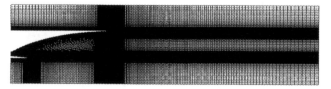

Figure 10.3 Typical grid in the nozzle.

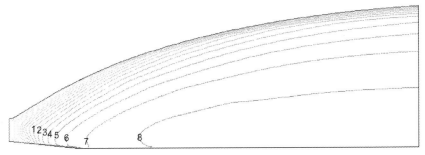

1) 2221.65   2) 2238.68   3) 2255.7   4) 2272.73   5) 2289.76   6) 2306.79   7) 2323.82   8) 2340.84

Figure 10.4 Total temperature contours of chemical nonequilibrium flow (H = 25 km, Ma = 6).

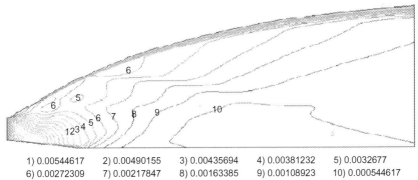

1) 0.00544617   2) 0.00490155   3) 0.00435694   4) 0.00381232   5) 0.0032677
6) 0.00272309   7) 0.00217847   8) 0.00163385   9) 0.00108923   10) 0.000544617

Figure 10.5 Mass fraction contours of OH in chemical nonequilibrium flow (H = 25 km, Ma = 6).

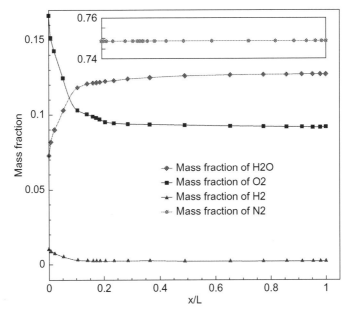

Figure 10.6 Mass fraction distributions near the wall (H = 25 km, Ma = 6).

decreases gradually, whereas the distribution of H_2O increases gradually and of N_2 does not change. The largest variations are close to the inlet of the nozzle. This means that the chemical reactions mainly appear close to the nozzle inlet.

The major conclusions from the study can be summarized as (1) the chemical reactions occurred mainly close to the nozzle inlet, (2) the inlet temperature and pressure affect the strength of the chemical reactions, and (3) close to the nozzle inlet, the mass fractions of some species varied significantly.

10.6.2 Shock Wave—Boundary Layer Interactions

Shock wave—boundary layer interactions (SBLIs) have significant interest in the aerospace community, as they occur in both external and internal aerodynamic problems at transonic, supersonic, and hypersonic speeds. There is a big interest in accurate numerical prediction methods using CFD. A review [25] presented the advances in understanding low-frequency unsteadiness, heat transfer prediction capability, phenomena in complex multishock boundary layer interactions, and flow control techniques. Several workshops have also been held through the years. Important conclusions are that in RANS predictions of SBLI the choice of turbulence model and the grid density are very important in enabling more accurate solutions.

10.7 CONCLUSIONS

Computational approaches for analysis of transport phenomena based on CFD were briefly summarized and reviews of recent works were provided. However, many challenges need to be overcome before CFD can take over as a primary tool in aerospace applications. The following issues need to be addressed: code and performance improvements; introduction of multiprocessor technology; improvements in grid technique including unstructured grids, adaptive grids, and deformable boundaries; validation and improvement of turbulence modeling; handling of boundary layer transition; and improvement of multiphase flow modeling. Besides, well-documented validation data bases are needed.

Results have revealed that the CFD approach, in many cases, can not only demonstrate important physical effects but also provide satisfactory results in decent agreement with corresponding experiments.

REFERENCES

[1] Sengupta TK. Theoretical and computational aerodynamics. Wiley; 2015.

[2] Patankar SV. Numerical heat transfer and fluid flow. New York: McGraw-Hill; 1980.

[3] Andersson DA, Tannehill JC, Pletcher RH. Computational fluid mechanics and heat transfer. 2nd ed. USA: Taylor and Francis; 1997.

[4] Smith GD. Numerical solution of partial differential equations. London: Oxford University Press; 1978.

[5] Reddy JN, Gartling DK. The finite element method in heat transfer and fluid dynamics. Boca Raton, Fla: CRC Press; 2010.

[6] Lewis RW, Morgan K, Thomas HR, Seetharamu KN. The finite element method in heat transfer analysis. J. Wiley and Sons; 1996.

[7] Kandelousi MS, Ganji DD. Hydrothermal analysis in engineering using control volume finite element method. Oxford: Academic Press; 2015.

[8] Wrobel LC. Boundary element method—volume 1 applications thermo-fluids and acoustics. UK: J. Wiley and Sons; 2002.

[9] Rhie CM, Chow WL. Numerical study of the turbulent flow past an Airfoil with trailing edge separation. AIAA J 1983;21:1525−32.

[10] Jang DS, Jetli R, Acharya S. Comparison of the PISO, SIMPLER and SIMPLEC algorithms for treatment of the pressure velocity coupling in steady flow problems. Numer Heat Transf 1986;10(3):209−28.

[11] Issa RI. Solution of the implicity discretized fluid flow equations by operator-splitting. J Comput Phys 1986;62:40−65.

[12] McBride D, Croft N, Cross M. Combined Vertex-based-Cell-Centred finite volume method for flow in complex geometries. In: Third International Conference on CFD in the minerals and process industries, 351-1356. Melbourne, Australia: CSIRO; 2003.

[13] Farhanieh B, Davidson L, Sunden B. Employment of the second-Moment closure for calculation of recirculating flows in complex geometries with collocated variable arrangement. Int J Numer Meth Fluids 1993;16, opp. 525−54.

[14] Pope S. Turbulent flows. Cambridge, UK: Cambridge University Press; 2000.

[15] Wilcox DC. Turbulence modeling for CFD. 2nd ed. La Canada, California: DCW Industries, Inc; 2002.

[16] Durbin PA, Shih TI-P. An Overview of turbulence modeling. In: Sunden B, Faghri M, editors. Modeling and simulation of turbulent heat transfer. Southampton, UK: WIT Press; 2005. p. 3−31.

[17] Spalart PR, Allmaras SR. One-equation turbulence model for aerodynamic flows. AIAA Paper 92-0439. 1992.

[18] Durbin PA. Separated flow components with k-ε-v2 model. AIAA J 1995;33(4):659−64.

[19] Menter FR. Zonal two-equation k-ω models for aerodynamic flows. AIAA Paper 93-2906. 1993.

[20] Launder BE. On the computation of convective heat transfer in complex turbulent flows. ASME J Heat Transf 1988;110:1112−28.

[21] Spalart PR, Jou WH, Stretlets M, Allmaras SR. Comments on the Feasibility of LES for Wings and the hybrid RANS/LES approach. In: Liu C, Liu Z, editors. Advances in DNS/LES. Columbus: Greyden Press; 1998.

[22] Spalart PR, Venkatakrishnan V. On the role of and challenges of CFD in aerospace industry. Aeronautical J 2016;120(1223):209−32.

[23] Maicke BA, Majdalani J. Evaluation of CFD codes for hypersonic flow modeling, AIAA 2010-7184. In: 46th AIAA/ASME/SAE/ASEE Joint Propulsion Conference & Exhibit; July 25−28, 2010.

[24] Wang XY, Xie GN, Sunden B. Analysis and calculation of chemical non-equilibrium turbulent flow in a scramjet nozzle. ASME GT2009-59638. 2009.
[25] Gaitonde DV. Progress in shock wave/boundary layer interactions. Prog Aerosp Sci 2015;72:80—90.

CHAPTER 11

Measuring Techniques

11.1 INTRODUCTION

Many physical properties in spacecraft are of vital concern, e.g., temperature, pressure, flow rate, and propellant mass gauging. The information of these parameters will form the important basis for the temperature and pressure control of the whole spacecraft, especially the payload and propulsion subsystem. Some of the parameter measurements in orbit can be achieved by referring to the methods used on the ground, with adaptive changes. For example, pressure transducers are applied to measure the pressure by satisfying the space working temperature. However, temperature measurement is affected because of the large temperature gradient caused by the absence of efficient convection heat transfer and exposure to the Sun and dark deep space. Liquid propellant will be stratified. Phase change occurs in the flow and tank because the flow rate and mass gauging are affected. Thus measurements of the temperature, flow rate, and mass are important. The compression mass gauge (CMG) method for mass propellant will be illustrated specifically.

11.2 TEMPERATURE MEASUREMENT

The instruments and equipment on spacecraft work in the required temperature range provided by the temperature system. However, the complexity of the spacecraft makes it not easy to establish an exact thermal control method. The experiments in orbit environment with vacuum and intense radiation data required modification of the thermal control system to provide suitable temperature condition for devices. Temperature and pressure are also used to determine the density of fluids and also to calculate the mass of the propellant. Also heat transfer properties require accurate temperature measurement.

As satellites orbit, one side is exposed to the direct radiation from the Sun, whereas the opposite side is completely dark and exposed to the deep cold outer space (3 K). This causes severe temperature gradients affecting the reliability and accuracy of the temperature. For cryogenic propellant

Heat Transfer in Aerospace Applications
ISBN 978-0-12-809760-1
http://dx.doi.org/10.1016/B978-0-12-809760-1.00011-9

storage, thermal stratification is formed without the efficient heat exchange of convection. Beduz et al. [1] performed wall heating stratification studies of liquid oxygen (LOX) and liquid nitrogen (LN_2) in small Dewar flasks. By using temperature measurements in the liquid, they were able to map the morphology of the temperature field. Below the interface, they observed a thin thermally conducting layer a few hundred microns thick with a steep temperature gradient, which resided on top of a convective layer with a shallow temperature gradient. Das et al. [2] used a dye injection system to map out the temperature field in a side-heated cavity containing water. This implies a big challenge for temperature measurement.

Temperature measurement devices should not introduce any change to the temperature field and should be compatible with the measured objects such as the solid and fluid. Temperature measurement on spacecraft requires that the sensor works stably for a long time under vacuum and ionic radiation. Because of the temperature gradient, the measurement needs several test ports or test arrays. The structure and size of the temperature sensor should satisfy the measured part with efficient heat transfer and measure the actual temperature with small error. Its weight and energy consumption should be considered carefully for a space mission. The test port array must be reasonable and can help understand the temperature distribution map. The measuring range covers the low temperature in cryogenic systems and high temperature with linear output.

The physical principles affected by temperature form the basis for temperature-measuring instruments. It is therefore reasonable to study temperature measurement by dividing the instruments used into separate classes according to the physical principle on which they operate. This will give several classes of instruments based on the following principles: (1) thermoelectric effect, (2) resistance change, (3) sensitivity of the semiconductor device, (4) radiative heat emission, (5) thermography, (6) thermal expansion, (7) sensitivity of fiber-optic devices, (8) color change, and (9) change of the state of the material. The latest development in temperature measurement is the introduction of microelectromechanical system (MEMS)-based devices. These are usually noncontact devices that are based on thermocouples or thermopile sensors (a thermopile is a number of thermocouples connected in series), with polysilicon gold thermocouples being the common choice. The sensor is powered by a battery that has a very long lifetime owing to the very low power consumption of the device.

Typical temperature sensors used on spacecraft are the thermocouple and the thermistor. The thermocouple is widely used on the ground experiment for its better stability, sensitivity, and low cost. The principle of a thermocouple requires the cold junction compensation at the reference temperature. The measurement accuracy is affected by the compensation end. The compensation is not easy to achieve on orbit. Furthermore, the electric signal is weak, which means it can be easily disturbed by the environment; the thermocouple wire can be damaged more easily than that of the thermistor; and its reliability has no advantage. Therefore, thermocouples are rarely applied on spacecraft in orbit. By contrast, thermistors have high measurement accuracy and reliability. But the cost is much higher and hysteresis may occur while measuring the temperature. As accuracy and reliability are the most important factors for a space mission and the temperature tends to be at a steady state, thermistors are a better choice for the flight test and space application. Noncontact methods, such as the use of infrared radiation thermometer, have also been studied [3]. But the measurement accuracy still needs additional research for improvement.

Both the temperature and the temperature gradient near a burning surface are important when modeling the combustion of solid propellants or hybrid rocket propellants because the temperature of the burning surface provides a critical boundary condition for models used to predict the regression rate of the propellant. The use of embedded micro-thermocouples has been a convenient and effective method for measuring temperature profiles throughout the thermal wave near the burning surface of such propellants. An inverted U-shaped thermocouple is better than an inverted V-shaped thermocouple to measure the temperature profile near the surface because there is less heat loss through the leads [4]. The effect of heat conduction through the thermocouple leads has been discussed in several studies [5—7]. These studies are valuable when discussing the effect of heat conduction through the leads. However, the spatial resolution of the thermocouples is inappropriate when compared with the thickness of the thermal wave. In addition, it is difficult to directly evaluate the experimental error due to the heat loss from the thermocouple leads. The effect of the thermocouple wire diameter on the burning surface's temperature measured by an inverted V-shaped thermocouple has not been discussed sufficiently [8]. Different thermocouple diameters were used to determine the mean evaporation rate, which was extrapolated by plotting the mean evaporation rate of the droplet as a function of the thermocouple diameter squared. By using thermocouples, it is also possible to obtain the fuel droplet temperature history, which shows, for butanol, that most of the evaporation process takes place at a constant droplet temperature [9].

Ishihara [10,11] evaluated the measurement error associated with an inverted V-shaped thermocouple in measuring the burning-surface temperature of a propellant. He also approximately analyzed the error in the measured surface temperature of a burning propellant caused by heat loss from the leads. In addition the effect of the initial temperature on the burning-surface temperature was experimentally examined.

In the conventional thermistor technique, the temperature rise within the thermistor is kept constant [12–15]. This requires a variation of the bead power dissipation during the measurement, based on an instantaneous step change [16]. The estimation of thermal conductivity is extracted from steady state, which requires collection of data for at least 5 s to enable extrapolation to infinity to get the steady state. This means that these data are affected by natural convection effects in the case of liquids, conduction along the lead wire, and the return of a thermal wave after it reaches the boundary of the sample's container. To avoid these problems, an improved version of the conventional thermistor technique was proposed [17,18]. Using this technique the researchers have succeeded in measuring the thermal conductivity of the liquid and powder bentonite (Kunigel V1) and the mixture bentonite–silica sand. The accuracy in both measurements was found to be good, and for the bentonite, the data correlated very well with all the best available correlations related to porous materials and mixtures. A newly proposed thermistor technique for the simultaneous estimation of thermal conductivity and thermal diffusivity was investigated numerically and experimentally for its accuracy and applicability [19,20].

Gas molecules are rarely present in microgravity environment in space, so convection can be neglected. Heat transfer is thus mainly by conduction and radiation. If the temperature sensor is installed improperly, thermal resistance will increase and result in unexpected measurement error. Theoretic and experimental results show that for solid temperature measurements the measurement error is directly proportional to the temperature difference between the environment and the measured surface on the ground, i.e., $\Delta T \propto T_g - T_o$, where T_g is the environment temperature and T_o is the measured surface temperature. This is because on the ground, convection is dominating mechanism of heat transfer. However, radiation will be the main heat transfer mechanism under microgravity. Accordingly, the error will be proportional to the difference between the environmental temperature to the power of n and the measured temperature to the power of n, i.e., $\Delta T \propto T_g^n - T_o^n$, where n is an experimental constant.

11.3 FLOW MEASUREMENT

Flow measurement is concerned with quantifying the rate of flow of materials. The material measured may be in a solid, liquid, or gaseous state. When the material is in a solid state, the flow can only be quantified as the mass flow rate. When the material is in a liquid or gaseous state, the flow can be quantified either as the mass flow rate or the volume flow rate, respectively. A flow measurement in terms of the mass flow rate is preferred if very accurate measurement is required. In aerospace applications, the flow material is the fluid and the density depends on temperature and pressure. The volume flow rate will then be considered.

The flow measurement is especially of interest for geostationary satellites using Liquid Apogee Engines (LAEs), where more than 80% of the propellant is consumed during apogee raising maneuvers. After orbit transfer, the remaining propellant is calculated with the "pulse count" (book keeping) method. The propellant consumption during station keeping maneuvers is calculated by recording all maneuver data (e.g., pulse duration, pulse mode, thruster temperature), making use of on-ground individual thruster calibration data. As the pulse count method starts evaluation at the beginning of satellite orbital life, the accuracy of the remaining propellant mass prediction is eventually driven and limited by the precision with which the quantity of propellant consumed during LAE firing is estimated. The flow meter gauging assessment for a typical large platform has resulted in an accuracy improvement compared to the traditional bookkeeping methodology. In addition to the improvements in propellant gauging, the flow measurements for each bipropellant could be used to minimize mixture ratio residuals by subsequent boost heat of the propellant tank during the operational life. Mixture ratio adjustment should result in lifetime extension. Furthermore, the flow measurement is a critical technique in the on-orbit resupply technology. It gives the resupply propellant mass information when the resupplying is in process.

11.3.1 Typical Flow Meters

Flow measurement research focuses on a single phase in the earlier study. Turbine flow meters are very reliable and commonly used to measure flow of valuable fluids [21]. The frequency at which the blades turn represents a given flow because each blade sweeps out a fixed volume of fluid. These fluids include natural gas, petroleum-based fuels, and hydraulic fluids. These devices have experienced many changes and improvements over the years since the first axial vanned flow meter was used by Reinhard Woltman in 1790 to measure water flow [22]. A survey of flow-measuring devices indicated that vortex shedding flow meters would also be able to measure hypergolic flows with the added advantages of no moving parts and no damage during high-volume-flow nitrogen purges [23]. An improved version of the magnetic flow meter burner developed at the ONERA Palaiseau Center combined with an acoustic analysis of the standing wave above the surface of the burning propellant strand to use velocity and pressure oscillation data taken from the surface of the propellant to as far as 1.2 cm above it, increasing the statistical confidence in the calculated acoustic admittance [24]. An ultrasonic flow meter has been developed and qualified by Bradford Engineering BV that provides a direct mass flow measurement during geostationary transfer, hence represents a gain in the global propellant gauging end accuracy. The design is currently undergoing delta qualification tests in the frame of an Alphabus contract [25]. An online measurement of gas flow rate and liquid flow rate in wet gas with one V-Cone throttle device is developed. The two-phase mass flow coefficient is employed to correct the deviation of the measurement device when it is used to measure wet gas. The coefficient linearly increases with the liquid densiometric Froude number and is affected by the gas densiometric Froude number and the ratio of gas density to liquid density [26].

11.3.2 Two-Phase Flow Measurements

Two-phase flows under microgravity conditions appear in a large number of important applications in fluid handling and storage and in spacecraft thermal and power systems (e.g., condensers, evaporators, piping system). The physics of this ubiquitous flow is, however, very complex and not well understood because of the absence of efficient heat transfer and the different flow rates of the liquid and gaseous phases. Since the early adiabatic two-phase microgravity experiments of Hepner et al. [27], researchers quickly realized the vast differences in interfacial behavior between terrestrial and reduced-gravity environments. Only three of the classical flow patterns in tubes are commonly achieved in reduced gravity: bubbly, slug, and annular flows, with a fourth frothy slug—annular flow pattern observed in a few studies, based on combinations of superficial velocities of vapor and liquid, respectively [28]. Flow measurement in the two-phase flow becomes a challenging investigation. In addition to the mass or volume rate, the velocity, void fraction, density, and pressure drop were also studied [29].

A few investigators explored pressure drop in microgravity flow boiling. Luciani et al. [30,31] performed parabolic flight experiments to investigate flow boiling of HFE-7100 (methoxyperfluorobutane) in rectangular channels. The microgravity and hypergravity pressure drop data from these experiments were also analyzed by Brutin et al. [32] and compared with the terrestrial $(1-g_e)$ data. They found that the two-phase frictional pressure drop increased with increasing gravity, which they attributed to an observed decrease in void fraction increasing the portion of the channel's cross-sectional area dedicated to liquid flow. Interestingly, this trend contradicts experimental findings from adiabatic two-phase flow studies [33–35], which found the two-phase frictional pressure drop to increase in microgravity, especially at low flow rates. Misawa [36] performed both drop tower and parabolic flight experiments to investigate flow boiling of R-113 (trichlorotrifluoroethane) in microgravity. Misawa used a square channel equipped with a heating film, and two electrically heated coiled tubes. The wall shear stress in microgravity was found to be 1.18 times larger than that in $1-g_e$, which they attributed to larger bubbles in the low-quality region in microgravity.

An experimental study of the film characteristics in annular flow was carried out by Jong and Gabriel [37] at a simulated microgravity condition experienced onboard NASA's DC-9 (Douglas Commercial-9) zero-gravity aircraft. The measurements of film thickness were taken during flight tests, along with high-speed video imaging, pressure drop, and void fraction data. The film properties were obtained from the film thickness time traces. From the film thickness traces, the average wave minimum height, the average film thickness, and the average wave height were calculated. The velocity and average frequency of the disturbance waves were also calculated using statistical techniques. The wave velocity is mostly unaffected by the reduction in gravity, except in regions of falling film flow at 1 g vertical flow. The wave frequency decreases with the reduction in gravity.

The gas–liquid two-phase flow should be studied and analyzed because it plays important roles in industrial applications. However, knowledge of two-phase flow and measuring techniques related to key parameters are not sufficient so far because the phenomena in two-phase flow are quite complex. There are many parameters in a two-phase flow. Among them, the void fraction, which is the ratio of the gaseous phase volume to the total control volume, is particularly significant. It is also a key physical parameter for determining other numerous key two-phase parameters including the two-phase flow density, the two-phase flow viscosity, and the average velocities of the two phases [38]. Moreover, the void fraction plays an important role in the modeling of two-phase flow regime transitions, heat transfer, and pressure drop. The knowledge of the void fraction is also crucial in many thermohydraulic simulations, such as coupled neutron—thermohydraulic calculations and two-phase natural circulation loop flow rates and heat transport rates prediction [39].

Existing measuring methods for void fraction include the differential pressure method, quick closing valves method [40], needle contact probe method [41], optical fiber probe method [42], conductance method [43,44], capacitance method [45], X-ray scanning method [46], gamma ray method [47], and neutron radiography [48]. In recent years, measurement using wire-mesh tomography has been studied [49,50]. Among these techniques, the quick closing valves method and the needle contact probe method are widely used because they help obtain the void fraction easily without extensive instruments. However, these measuring methods have some problems. For example, the quick closing valves method requires stopping the flow, and the measured value of void fraction often includes large errors in the high void fraction region. The needle contact probe method needs a sufficiently thin needle (measuring several micrometers) as a sensing element; however, making very thin needles is not easy.

As mentioned earlier, establishment of a new simple measuring method for void fraction has important consequences. A study proposed a new measurement technique using a fully diluted magnetic fluid. The magnetic fluid is a magnetized liquid that has many applications [51]. The measuring method is based on the idea that there is a relationship between the space distribution of air bubbles and the distribution of the magnetization within a diluted magnetic fluid. The proposed measuring technique does not require any pipe processes and can be realized as an easy measurement of void fraction because the measurement is carried out using electromagnetic induction. The measuring technique is also available for applications using gas—liquid two-phase flow in pure and mixtures of magnetic fluids [52,53].

11.3.3 Microscale Fluid Flow Measurement

Vigorous studies are underway to understand the movement of a microscale fluid. Especially, the understanding of gas—liquid two-phase flow characteristics in a microchannel is essential for developing and designing microdevices such as microreactors, mobile-type fuel cells, and micro heat exchangers. Concerning the flow characteristics, Serizawa et al. [54],

Kawahara et al. [29,55–60], Chung and Kawaji [61,62], and Kawaji et al. [63] reported unique differences between the microchannel and the conventional sized channel. The two-phase frictional pressure drop and the void fraction correlations from literature were tested against their data. Furthermore, for the void fraction, an analytical code based on a steady-state adiabatic one-dimensional two-fluid model was also tested.

11.4 LIQUID MASS GAUGING IN MICROGRAVITY

11.4.1 Review

In low gravity, the position of a liquid in a container may be markedly different from that on the Earth, where the liquid occupies the bottom of the container and forms a horizontal free liquid–gas surface at the lowest possible level within the container. Also, in low gravity, liquid and gas are not dominated by the strong buoyancy force of the Earth's gravity, fluid may become a mixture of gas bubbles of many sizes interspersed within a liquid, and liquid may not be at either the "bottom" or the "top" of the container. Consequently the familiar gauging methods used on the Earth are not generally applicable in space.

As the available gauging methods do not work in low gravity, there is a strong need for quantitative gauging systems for future space missions in which subcritical liquid fluids, particularly cryogenic liquids, will be stored and handled. Typical gauging applications include liquids for life subsistence and engine propellants. Liquid mass gauging is also required for fluid resupply to on-orbit systems for vehicles ranging from spacecraft to large space platforms. Future applications include propellants and life support fluids for the space station alpha [64] on interplanetary propulsion systems, and manned lunar vehicles.

The two prevalent techniques for quantifying the liquid level in a spacecraft tank are the bookkeeping method and the PVT (pressure, volume, temperature) method. In the bookkeeping method, the signal from a flow meter on the tank outlet tubing is recorded and integrated to estimate the total quantity of liquid removed from the tank. This method is simple, but the measurement errors increase with time as the tank is drained and special care must be taken to avoid two-phase flow at the meter. In the PVT method, noncondensable gas from a high-pressure auxiliary tank is injected into the main liquid tank. The responses of both the pressurized tank and the main tank to this mass transfer between the tanks are used to compute the vapor volume in the main tank, providing an estimate of the liquid volume. The PVT method requires the use of high-pressure tanks, valves, and multiple temperature and pressure sensors in both tanks. The combination of errors associated with the sensor suite degrades and complicates specification of the overall uncertainty estimate.

Many "low-g quantity gauges" have been investigated in concept or tested in laboratory environments over the past 50 years. These systems have been based on the use of a variety of physical principles such as radio frequency microwaves [65–67], gas bubble resonant frequency, liquid heat capacity [68–71], optical absorbency [72–74], ultrasonics [75–77], capacitance [78,79], acoustics [80], gamma ray densitometry [81], and flow meters for monitoring liquids leaving and entering the tank [25,75,76]. However, they have all proved to have significant limitations in gauging accuracy, complexity, or weight. The CMG [82] method of adiabatic compression proposed by the Southwest Research Institute has been the one in focus of significant efforts.

11.4.2 Compression of Mass Gauging

In the CMG method, a slight volume change is applied to a tank so that the pressure changes to conform to the volume adjustment. The volume-changing device is flexible and can extend and shrink, as shown schematically in Fig. 11.1. Transducers are employed to measure the tank static pressure and dynamic pressure. By assuming that the mixture behaves as an ideal gas, the tank walls to be rigid, the compression to be adiabatic, and the change in the tank volume $\Delta V_t \ll V_g$, then the vapor volume V_g is given by the thermodynamic relation [82]

$$V_g = \gamma_0 P \frac{\Delta V_t}{\Delta P} \tag{11.1}$$

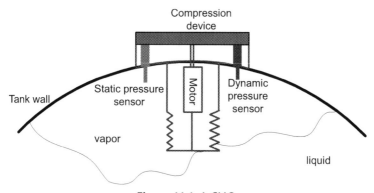

Figure 11.1 A CMG.

where γ_0, P, ΔV_t, and ΔP are the ratio of specific heats for the gas, the tank static pressure, and the amplitudes of the oscillating tank volume and the pressure, respectively. All these properties are determined using absolute values, except where stated specifically.

The tank total volume V_t is calibrated before the experiment. The liquid volume is obtained by subtracting the gas volume computed in Eq. (11.1) from the tank total volume. The liquid mass is then determined using the volume multiplied by the propellant density ρ_l:

$$m_l = \rho_l V_l = \rho_l (V_t - V_g) \tag{11.2}$$

Mord et al. [82] found that the ideal model may be influenced by heat transfer at the liquid—gas interface. An empirical constant C needs to be employed to eliminate this effect. The tank compressibility is corrected by a correction factor α, which is the tank stiffness factor. Rogers et al. [83] clarified that the actual vapor volume change seen in Eqn (11.1) should be modified by an effective change that considers heat and mass transfer between the two phases. More laboratory tests were conducted to investigate the CMG performance, especially in cryogenic environments. However, many problems appeared, such as erratic pressure measurements, the influence of the test facility configuration, and low-frequency signals. These were described by Jurns and Rogers [84] and Dodge and Kuhl [85]. Green et al. [86] developed a flight-like CMG design and conducted ground tests. Technologic development has advanced the CMG maturity through alcohol and LN_2 tests. Location effects on the CMG and pressure sensors were studied and a frequency dependence on the calculated gas volume was observed for different locations. However, additional tests were required to verify the frequency dependence phenomenon and its cause. Fu et al. [87] established a CMG ground test system and verified the critical components for gauging. The experimental results showed that the measurement accuracy was within $\pm 1\%$. Harmonic influence was identified at different compression frequencies and some steps were recommended to reduce its impact to obtain relatively exact pressure change amplitudes.

In addition to the laboratory tests described earlier, the NASA Lewis Research Center (LeRC) used Learjet flights to examine CMG performance in microgravity. The results were not satisfactory because of the noise and vibrations from the Learjet. Monti and Berry [88] described a CMG experiment conducted by the European Space Agency in the middeck of a space shuttle with test fluids such as Fluorinert and gaseous helium. The test results were generally in agreement with expectations because of the excessively large ratio of bellow displacement volume to tank volume.

Based on the previous work, further research to determine other influencing factors on the CMG performance is required before implementing it in the field. The main focus here will be to describe an experimental validation of the CMG performance under attitude disturbances and in real thermal environments.

11.4.2.1 Description of Ground Experiments

Three types of experiments were conducted to verify the performance of the CMG method. First the CMG method was manipulated without any external disturbance. Then attitude disturbances and heat fluxes were employed to simulate measurements in a real space environment. All the experiments shared the same main components contained in the first experimental setup. A ground breadboard test system was developed and fabricated in the laboratory. The main components of the system were as follows: (1) tank prototype, (2) compression and drive motor assembly, (3) transducers, (4) data acquisition equipment, and (5) calibration subsystem. In addition, vibration mechanisms and heaters were developed. For each set of experiments, measurements were conducted at different fill levels covering both high- and low-liquid-propellant proportions in a tank. Different compression frequencies within 10 Hz were applied at every fill level. The fluids tested were water and air.

11.4.2.1.1 Experimental Apparatus

Fig. 11.2 shows the schematic diagram of the experiment. The dimensions of the cylindrical tank are 0.3 m diameter, 0.4 m height, and 28.2 L internal volume. Sealing properties were considered and leakage tests were conducted after all the subsystems were installed. No leakages were detected. The compression device consisted of a direct current motor, which was enclosed in a stainless steel canister, coupled by means of an eccentric drive shaft to a stainless steel bellow attached to the container. The rotation of the motor drive shaft was converted to a liner motion to operate the bellow. The eccentricity of the drive shaft was designed in such a way that the linear displacement of the bellow could be varied. For these tests, the stroke was set to result in a total ΔV_t of 17.21 cm^3. The ratio of the displacement ΔV_t to test tank volume was in the order of 10^{-4}, which is typical for compressibility gauging devices [82]. The motor speed was impulse controlled, and the pulse frequency could be modified over wide ranges according to the test requirements.

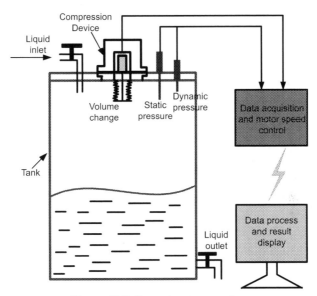

Figure 11.2 Ground test apparatus.

Two types of pressure transducers were used to measure the average pressure and the oscillating pressure amplitude of the tank. The first transducer was an absolute pressure sensor with a full-scale accuracy of $\pm 0.05\%$, which was used to sense the environmental pressure and the tank static pressure. The second transducer was a differential pressure sensor with the same accuracy. It was used to measure the pressure difference between the tank and the ambient environment. The transducer had a measurement range of ± 7 kPa. To provide a reference for the CMG accuracy evaluation, a high-accuracy bench scale was developed to calibrate the CMG. During testing, the weight of the empty tank and the partly filled tank was measured to determine the amount of liquid required for various fill levels. This method provides accurate liquid mass measurement for a known condition before operating the CMG. By comparing the measured and calibrated masses, the gauging error could be obtained. Data acquisition was controlled by a PC, with software written in the C programming language.

11.4.2.1.2 Experimental Procedures
In the first set of experiments, no external disturbance was applied to the experimental system. Measurements were carried out at the fill levels of 0%, 10%, 30%, 50%, 70%, and 90% of the tank volume. For each fill level, the

compression frequency of the compression device was changed from 1 to 5 Hz. The stability of the experimental system was checked and the measurement accuracy was obtained. These results were compared with those taken under external excitation, such as changing the thermal environment considering the heat transfer between the tank and ambient and the tank-induced vibration from an attitude disturbance.

A schematic of the test setup used to introduce attitude disturbance is shown in Fig. 11.3. The test tank and assembled components were vibrated on a cart using a linear motion actuator. The vibration amplitude and frequency were controlled using preset programs. Vibration frequencies close to the natural frequency of the partially filled tank were chosen for this investigation. Sinusoidal vibration disturbances in the horizontal direction

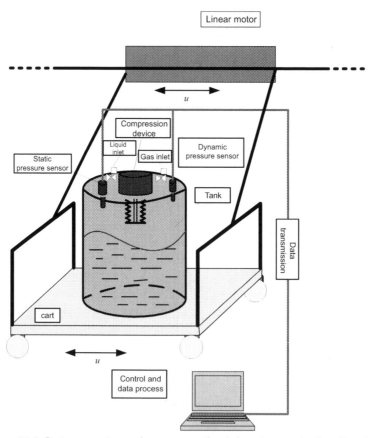

Figure 11.3 Basic experimental system vibrated using attitude disturbance equipment.

were performed using the linear motion actuator. Random rotation disturbances were irregularly applied to the cart to simulate a real space environment. Measurements were conducted at fill levels of 10% and 70%, which represent the respective low and high propellant percentages in the tank. Table 11.1 shows the experimental matrix for the attitude disturbance tests.

Another series of tests were conducted to investigate the effects of creating thermal stratification in the liquid as the measurements were conducted and of cooling the heated liquid by reducing the ambient temperature. The schematic diagram for these setups is shown in Fig. 11.4. Normally, phase change occurs when heat is transferred into or out of the tank. By doing this, the influence of the thermal environment on the liquid mass gauging accuracy can be examined. Measurements at every 20% of the fill level, ranging from 10% to 90%, were conducted and analyzed.

11.4.2.2 Test Results and Discussion
11.4.2.2.1 Normal Tests
Table 11.2 shows the CMG experimental results for six different liquid fill fractions at various compression frequencies. The tests were repeated for each case. The reference mass is calibrated using the high-accuracy bench scale. This demonstrates that the data stabilize around a certain value, with a discrepancy from the reference mass at every fill level. The discrepancy decreases as the fill level increases and is almost identical when the fill level is above 90%. This indicates that the measurement accuracy decreases as

Table 11.1 Test Matrix for the Attitude Disturbance Tests

10%		70%	
Compression Frequency (Hz)	Sloshing Frequency (Hz)	Compression Frequency (Hz)	Sloshing Frequency (Hz)
1	0.5	1	0.5
	1		1
	1.5		1.5
1.5	0.5	1.5	1
	1		1.5
	1.5		
2	0.5	2	1
	1		1.5
	1.5		2

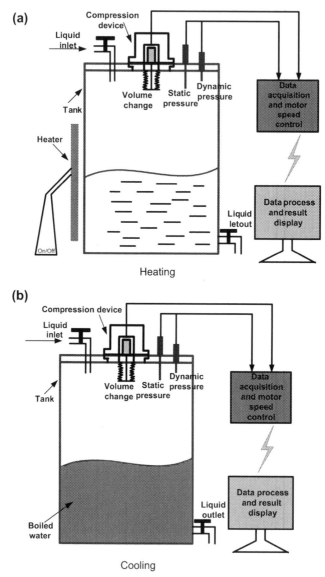

Figure 11.4 CMG thermal environment ground testing using (a) heating and (b) cooling.

time elapses because more propellant is being consumed by the engine. This is explained by the gauging uncertainty for the liquid fill level:

$$\frac{U_{V_l}}{V_t} = \left(\frac{U_{V_t}^2}{V_t^2} + (1-F)^2 \left(\left(\frac{U_P}{P}\right)^2 + \left(\frac{U_{\Delta V_t}}{\Delta V_t}\right)^2 + \left(\frac{U_{\Delta P}}{\Delta P}\right)^2 \right) \right)^{1/2} \quad (11.3)$$

Table 11.2 Test Results Under Normal Conditions

Compression Frequency (Hz)	Test Times	0	10% 2.8162 kg	30% 8.4485 kg	50% 14.0809 kg	70% 19.7132 kg	90% 25.3455 kg
1	1	1.246027588	4.038535371	9.34558764	14.68622875	19.98829054	25.32522342
	2	1.235247598	4.044861304	9.369243321	14.6556234	19.99531737	25.31360102
	3	1.261849091	3.987665958	9.374954796	14.60975542	19.99687839	25.32157997
	4	1.220757012	4.072741104	9.341861035	14.66161487	19.9870354	25.32504247
	5	1.247141711	4.022046515	9.354087291	14.67391774	19.98381991	25.33579427
	6	1.18366578	4.094527218	9.399746272	14.67613203	20.00175439	25.32592177
2	1	1.195580236	4.099113138	9.422276634	14.74570737	20.01889643	25.35781289
	2	1.233371838	4.165488409	9.382663142	14.73127778	20.05762689	25.3685424
	3	1.211002781	4.227565515	9.5662464	14.71699046	20.04741142	25.36844084
	4	1.280704203	3.925026838	9.393471761	14.74536884	20.0416847	25.367097
	5	1.269368139	4.084387591	9.524568538	14.74079945	20.05273038	25.3671058
	6	1.357979959	4.088029111	9.45048334	14.73262975	20.04222127	25.3734082

where U is the uncertainty of the measured or computed parameter value and F is the fill level. At low fill levels, the large gas volume is less sensitive to the same volume change compared with that at high fill levels. Therefore, the gauging error increases as the gas volume increases.

Although the measured results do not match the reference data, the test system functioned well. The deviation between the measured and reference values resulted from the systematic error and can be mitigated with multiple metering. The gauge accuracy is obtained in Table 11.3 after elimination of the systematic error. The gauging error is typically within ± 1% of the full-tank propellant volume and is well within the uncertainty requirement derived for the target flight application. Furthermore, at lower compression frequencies, the measured values are below the "real" mass, whereas at higher compression frequencies, they are above the "real" mass. The gauging error at low fill levels is larger than that at high fill levels for any compression frequency.

Fig. 11.5 shows the comparison between the experimental data for different fill levels. Fig. 11.5(a) shows that a linear relation is observed between the measured and calibrated liquid mass. Ideally a line with a slope of unity should be obtained for this comparison. Ideal calibration is observed when the liquid mass is above 15 kg (above the 50% fill level). The error increases greatly below the 50% fill level, especially as the filled volume approaches zero. This is because of the systematic error described previously. The desired measurement accuracy can be achieved if the systematic error is determined for enough fill levels by multiple metering or by setting up a relationship between the measured and the calibrated masses, as in Fig. 11.5(a). The measurement error trends in Fig. 11.5(b) illustrate that the measured liquid mass is higher than the reference values, or real values, for high compression frequencies, whereas it is less than the real values at low compression frequencies. This is because more noise is generated at high compression frequencies that contribute to the pressure change amplitude, resulting in a low measured gas volume.

11.4.2.2.2 Attitude Disturbance Tests

Table 11.4 shows the experimental results obtained under sinusoidal vibration for two liquid fractions. A sloshing frequency of 0 Hz means there is no vibration. Two test times were selected for every oscillation test. The disturbance frequencies were around the natural frequency of the partially filled tank, approximately 1.181 Hz for a 10% fill level and 1.753 Hz for a 70% fill level. Additionally the sloshing amplitude was small. The average

Table 11.3 Gauge Accuracy Under Normal Conditions

Compression Frequency (Hz)	Test Times	0	10% 2.8162 kg	30% 8.4485 kg	50% 14.0809 kg	70% 19.7132 kg	90% 25.3455 kg
1	1	0.003	−0.355	−0.361	−0.133	−0.213	−0.131
	2	−0.035	−0.332	−0.277	−0.242	−0.188	−0.172
	3	0.059	−0.535	−0.257	−0.405	−0.183	−0.144
	4	−0.087	−0.233	−0.375	−0.220	−0.218	−0.131
	5	0.007	−0.413	−0.331	−0.177	−0.229	−0.093
	6	−0.219	−0.156	−0.169	−0.169	−0.166	−0.128
2	1	−0.176	−0.139	−0.089	0.078	−0.105	−0.015
	2	−0.042	0.096	−0.230	0.027	0.033	0.023
	3	−0.122	0.317	0.422	−0.024	−0.004	0.023
	4	0.126	−0.758	−0.191	0.077	−0.024	0.020
	5	0.086	−0.192	0.274	0.061	0.015	0.018
	6	0.400	−0.179	0.011	0.032	−0.022	0.040

All values are in percentage.

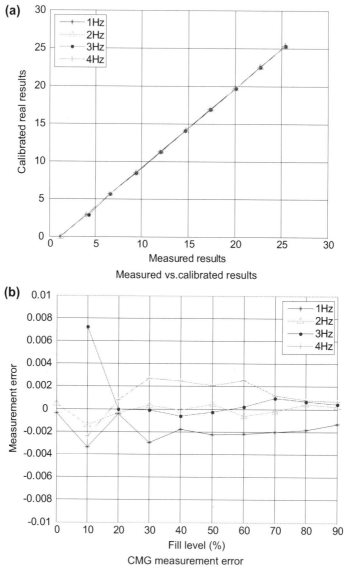

Figure 11.5 Measurement errors at various frequencies: (a) measured versus calibrated results and (b) CMG measurement error.

shaking velocity was approximately 0.08 m/s. The measurement accuracy was determined to be within ± 1% when the system was vibrated under a certain frequency that meets the gauging requirements. The natural frequency of the tank system is not easily excited when the amplitude of the

Table 11.4 Sinusoidal Vibration Test Results for Fill Levels of 10% and 70%

Compression Frequency (Hz)	Sloshing Frequency (Hz)	Test Times	Measurement Errors (%)	
			10%	70%
1	0	1	−0.692	−0.005
		2	−0.415	0.052
	0.5	1	0.957	0.043
		2	0.471	−0.033
	1	1	−0.964	0.476
		2	−0.697	0.033
	1.5	1	0.199	−0.879
		2	0.180	−0.551
1.5	0	1	0.196	−0.112
		2	0.367	−0.087
	0.5	1	0.267	
		2	0.580	
	1	1	−0.149	−0.461
		2	−0.230	−0.505
	1.5	1	0.082	−0.825
		2	0.591	−0.624
2	0	1	−0.139	−0.041
		2	0.096	0.097
	0.5	1	0.513	
		2	0.243	
	1	1	0.825	−0.172
		2	0.418	0.115
	1.5	1	−0.505	0.157
		2	−0.538	0.477
	2	1		−0.753

external attitude disturbance is low. In such cases, the influence of liquid sloshing on the propellant mass measurement cannot be ignored.

However, compared to the normal test results, the experimental results for the attitude disturbance tests had larger errors. For a 10% fill level and compression and vibration frequencies of 1 Hz, the measurement error increases sharply to almost ± 1%. This is a result of the three frequencies (compression, natural, and vibration) being quite close and resulting in surface waves in the liquid and more noise signals. If the amplitude increases, the error may exceed ± 1% in this situation. When the compression frequency increases to 1.5 Hz, the measurement accuracy becomes acceptable again. The useful signal is 1.5 Hz, even though the vibration frequency is 1 Hz, which is close to the natural frequency of approximately 1.181 Hz. The noise generated can be filtered during the data analysis process so that the accuracy is kept at acceptable levels. The same phenomenon is found for a 2-Hz compression frequency.

For a 70% fill level, the measurement error is larger than that for the normal tests at vibration frequencies of 1.5 Hz and 2 Hz for any different compression frequency, which is close to the natural frequency of 1.753 Hz for a 70% filled tank. At high liquid fractions, the small gas volume is more sensitive to the noise induced by the liquid sloshing. The liquid sloshing causes the gas and liquid components to flow. This fluid flow is complicated and dynamic. The combined effects of fluid flow and noise contributes to large errors in the mass gauging. Therefore, the larger the liquid fraction in a mixture, the more the error due to the high sensitivity of the ratio of the small gas volume to the fluid flow. Predicting the natural frequency is needed to choose a compression frequency that avoids resonance. To ensure reliable gauging, the compression frequency is usually set to be less than the lowest natural liquid sloshing frequency.

Tests were also conducted when the test system was disturbed under random sloshing. The data from these tests are displayed in Table 11.5. The

Table 11.5 Random Vibration Tests Results

10%	Compression Frequency (Hz)	1 Hz	1-Hz Sloshing Frequency	2 Hz	2 Hz Sloshing Frequency
(a) 10% Fill level					
Test	1	0.238	0.431	0.475	0.060
times	2	0.478	0.147	0.440	0.351
	3	0.320	0.528	−0.072	0.250
	4	0.492	0.299	0.202	0.113
	5	0.663	0.120	0.257	0.256
	6	0.373	−0.073	0.401	0.463

70%	Compression Frequency (Hz)	1 Hz	1-Hz Sloshing Frequency	1 Hz Higher Speed
(b) 70% Fill level				
Test	1	−0.008	−0.012	0.035
times	2	0.108	0.022	0.073
	3	0.078	−0.020	0.058
	4	0.055	0.029	0.008
	5	0.025	0.080	0.035
	6	−0.008	0.002	0.057

All values are in percentage.

results show that the test system behaved well and the effect of random vibrations was not pronounced.

Normally the effect of attitude disturbance on a propellant mass gauge is observed by liquid sloshing. The resonance in sloshing, which occurs when the frequency of the tank motion approaches the natural frequency of the liquid inside, may cause large structural loads to develop in the tank frame. This resonance phenomenon may be connected to complex motions in the liquid that can couple with structural motions and represent a threat to the stability of the structure [89]. Liquid sloshing also presents a challenge to proper spacecraft attitude control [90,91]. External excitations cause waves to form in the liquid and the fluid to flow. The fluid imposes a force on the tank wall. The maximum impact force should be evaluated to ensure the tolerance of the pressure sensor and should be a consideration during transducer installation. If waves are excited at the same frequency as the useful signal, then data processing may become a difficult task. However, no valid results could be obtained for this condition. As described by Mord et al. [82], for a transducer requiring tubing, if liquid flows into the tube because of sloshing, the results do not need to be corrected. Ground testing shows that it is not very easy to excite a liquid's natural frequency. Even if the liquid in a tank is actually forced to create waves at the surface that are not synchronized with the compression, it is still possible to operate the CMG with high accuracy. Therefore, if the tank liquid's natural frequency can be ensured, it is better to implement the gauge and avoid exciting at that particular frequency.

11.4.2.2.3 Heat Transfer Tests

To investigate the effects of heat leakage on the CMG, the tank sidewall was heated when performing the mass measurements. In addition, heat transfer from the tank is simulated by filling the tank with preheated water to create a temperature gradient between the tank and the ambient environment. Experimental data taken with heat transfer through the tank wall are displayed in Table 11.6. The measurement accuracy was found to satisfy the flight requirement. The gauge accuracy decreases but is still within the acceptable range of less than 1% error. This phenomenon is more obvious at lower fill levels. The influence increases with a decrease in the tank fill level because of the small heat capacity of the whole tank. For this case, both heat and mass transfer occur more easily.

The thermodynamic state, i.e., the pressure and temperature, of the fluid changes quickly. As a result, errors associated with the thermodynamic

Table 11.6 Results of Heat Transfer Tests

Compression Frequency (Hz)	Test Times	10% 2.8162 kg	30% 8.4485 kg	50% 14.0809 kg	70% 19.7132 kg	90% 25.3455 kg
(a) Liquid temperature higher than environment temperature						
1	1	0.031	0.410	0.054	−0.121	−0.082
	2	−0.469	0.446	0.070	−0.024	−0.087
	3	−0.172	0.198	0.126	−0.071	−0.077
	4	−0.450	−0.057	−0.033	−0.104	−0.090
	5	−0.201	−0.107	0.053	−0.072	−0.079
	6	−0.258	0.060	−0.284	−0.074	−0.240
2	1	−0.186	0.036	0.063	0.082	0.043
	2	−0.281	−0.111	−0.156	0.072	0.066
	3	0.281	−0.025	0.004	0.077	−0.714
	4	0.230	−0.061	−0.036	0.055	0.056
	5	0.240	−0.042	−0.032	−0.023	0.050
	6	−0.084	−0.098	−0.101	0.042	0.063
3	1	0.585	−0.394	0.127	−0.044	0.090
	2	0.256	−0.498	−0.076	0.314	0.075
	3	0.641	−0.544	0.433	−0.221	0.049
	4	0.709	−0.157	−0.115	−0.192	0.036
	5	−0.450	−0.422	−0.217	−0.091	0.165
	6	−0.268	−0.155	0.086	0.034	0.059
4	1	−0.340	0.087	−0.464	0.062	0.074
	2	0.400	0.557	−0.155	−0.214	0.095
	3	−0.260	0.429	0.370	0.252	0.096
	4	−0.100	0.058	0.121	0.273	0.077
	5	−0.187	0.402	0.021	−0.001	0.134
	6	0.334	−0.012	0.140	−0.011	0.140
(b) Measurements under heating						
1	1	0.096	−0.295	−0.162	−0.014	−0.160
	2	0.329	−0.141	−0.163	−0.031	−0.045
	3	0.326	−0.201	−0.027	−0.091	−0.080
	4	0.216	0.039	0.161	0.033	−0.051
	5	−0.182	−0.028	0.009	−0.012	−0.058
	6	−0.004	0.033	0.068	0.085	−0.015
2	1	−0.060	0.724	0.221	0.293	0.120
	2	0.564	0.147	0.341	0.265	0.109
	3	0.798	0.381	0.416	0.312	0.110
	4	0.374	0.476	0.289	0.268	0.077
	5	0.098	0.365	0.467	0.388	0.123
	6	−0.613	0.436	0.569	0.292	0.093

Table 11.6 Results of Heat Transfer Tests—cont'd

Compression Frequency (Hz)	Test Times	10% 2.8162 kg	30% 8.4485 kg	50% 14.0809 kg	70% 19.7132 kg	90% 25.3455 kg
3	1	−0.518	0.124	0.947	0.212	0.134
	2	0.579	0.365	0.791	0.574	0.107
	3	0.691	0.723	0.791	0.412	0.243
	4	−0.959	0.8647	0.861	0.571	0.197
	5	−0.863	0.190	0.924	0.449	0.089
	6	−0.399	0.743	0.846	0.403	0.123
4	1	−0.657	0.5136	0.423	0.238	0.264
	2	0.943	0.567	0.694	0.445	0.161
	3	0.286	0.595	0.609	0.653	0.203
	4	0.471	0.659	0.493	0.721	0.172
	5	−0.841	0.382	0.897	0.660	0.131
	6	−0.678	−0.005	0.223	0.649	0.220

All values are in percentage.

property changes are induced if the parameters used in the computational equation do not describe the local thermodynamic state.

Fig. 11.6 shows the comparison between the experimental data from the normal tests and those from the heat flux condition. The gauging error is acceptable under the different thermal environments. The tendency is that the measured values are larger under heating than those at normal conditions and are smaller under cooling conditions. This is very apparent at low fill levels. At high fill levels, the gauging error changes slightly under different heat transfer conditions compared to the normal tests.

As a result of the heat addition into the liquid and vapor phases, the internal energy of both phases increases and consequently, the tank pressure and temperature rise. If the liquid and vapor are sufficiently well mixed to ensure homogeneity within the system, the temperature in the tank is assumed to be uniform and liquid/vapor thermodynamic quantities are assumed spatially invariant and at saturation within each bulk phase. Neglecting the viscous dissipation and the rate of increase in temperature predicted by the first law of thermodynamics [92]:

$$\frac{dT_{sat}}{dt}\left\{\rho_v^{sat}V_v\frac{de_v^{sat}}{dT_{sat}} + \rho_l^{sat}V_l\frac{de_l^{sat}}{dT_{sat}} + \frac{d\left(\rho_v^{sat}V_v\right)}{dT_{sat}}L - \frac{dp_{sat}}{dT_{sat}}V_t\right\} = \dot{Q}_w \quad (11.4)$$

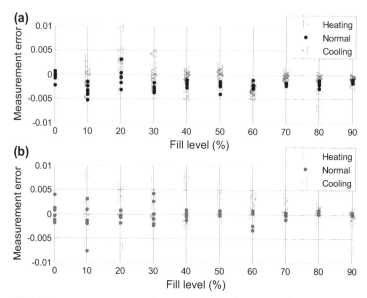

Figure 11.6 Measurement errors observed during heat transfer: (a) measurement errors associated with 1-Hz compression frequency and (b) those associated with 2-Hz compression frequency.

where the super-/subscript sat denotes an equilibrium state along the saturation line. The subscripts v and l represent the liquid vapor and liquid propellant, respectively; $\dot{Q}_w$ is the net heat entering the system through the tank wall; and L is the latent heat of evaporation. If a noncondensable gas is present as part of the vapor component, the above-mentioned model is modified to be

$$\frac{dT_{sat}}{dt}\left\{\rho_l^{sat}V_l\frac{de_l^{sat}}{dT_{sat}} + \rho_v^{sat}V_v\frac{de_v^{sat}}{dT_{sat}} + \rho_n V_v c_{pn} + \frac{d\left(\rho_v^{sat}V_v\right)}{dT_{sat}}L - \frac{dp}{dT_{sat}}V_t\right\} = \dot{Q}_w$$

(11.5)

where ρ_n and c_{pn} are the density and specific heat at constant pressure of the noncondensable gas, respectively. The gas pressure is the sum of the liquid, vapor, and noncondensable gas pressures $p = p_{sat} + p_n$.

For the ground experiments, liquid water is used as the propellant and air is used as the noncondensable gas. The pressure rise in the tank under a sidewall heating rate of $\dot{Q}_w = 50W$ is displayed in Fig. 11.7. The CMG operates for less than 2 min. The pressure rises gradually during this time. The rise rate decreases as the fill level increases. For a fixed heat load, as the liquid volume increases and because of the increasing fill level, the pressure

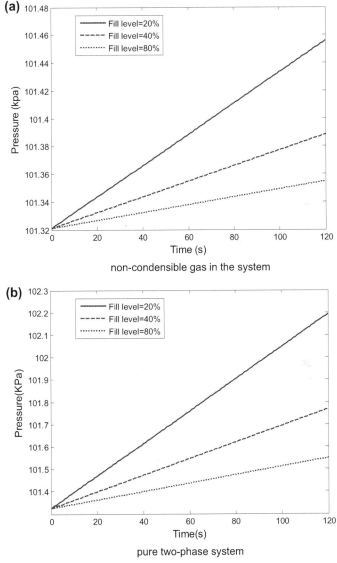

Figure 11.7 Pressure rise for different fill levels of water: (a) noncondensable gas in the system and (b) pure two-phase system.

rise rate is reduced. As seen in Fig. 11.7(a), for a 20% fill level, the greatest pressure rise is less than 150 Pa, which is 0.15% of the initial pressure. The saturation temperature changes by less than 0.5 K, which means density changes may be neglected for liquid mass computations. Thus high accuracy can still be achieved for both the volume and mass computations under a heat load. By comparing Figs. 11.7(a) and (b), the pressure rise is found to be reduced by the presence of noncondensable gas. This is because the noncondensable gas suppresses the phase change. Heat leakage is more common for cryogenic liquids, such as liquid hydrogen and LOX. Although the storage tanks are equipped with high-quality insulation, heat enters into the tank through conduction in the support structure and the pipelines that connect the tank to other devices. Therefore, the thermodynamics of the liquid hydrogen tank are estimated using the same heat load, as shown in Fig. 11.8. Helium is used as the noncondensable gas. The effect of the noncondensable gas can also be identified by comparing Fig. 11.8(a) and Fig. 11.8(b). However, the rate of the pressure rise increases by three or four orders of magnitude at the same fill level and is determined by the properties of the propellant. When using water, the bracketed portions of Eq. (11.4) or Eq. (11.5) are much larger than those for liquid hydrogen. The change in the gas ratio of specific heats γ_0 with temperature is approximately $3-10\%$ per degree at the saturation state, and -1.5% per degree for the liquid density. Thus in this case, the effective method for achieving high accuracy with the CMG is to guarantee high-quality tank insulation before operating the CMG. If heat leakage occurs, introducing a noncondensable gas in the ullage space is a method that can be used to control the pressure and temperature in the tank before the CMG is operated. Venting vapor for a short time can also cool the heated propellant.

If the liquid and vapor do not mix sufficiently well, thermal stratification occurs. The pressure rise in the initial transient period is different from the prediction made in the homogeneous model because of the convection induced by buoyancy. Self-pressurization and thermal stratification in a cryogenic tank requires further research.

11.4.3 Summary and Concluding Remarks

The CMG method was discussed in detail and an investigation using ground experiments was described. Based on the results, it appears that using a compression gauge is a suitable method for gauging liquids in low-gravity environments. A summary of the test results is as follows:
1. A gauging accuracy of $\pm 1\%$ was repeatedly demonstrated for different fill levels at various compression frequencies. Provided the gauging system is highly stable, the systematic errors can be mitigated by multiple

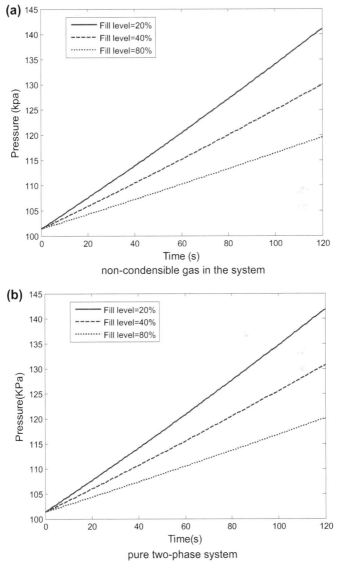

Figure 11.8 Pressure rise for different fill levels of liquid hydrogen: (a) noncondensable gas in the system and (b) pure two-phase system.

metering. The relation between the measured liquid mass and the "real" values can be established before using CMG. Therefore, for every measured value obtained by CMG, the "real" mass can be estimated through interpolation calculations using the calibrated relation.

2. When the ground test system is vibrated at several sloshing frequencies, the CMG can still be operated with acceptable accuracy as long as the transducers are located within the volume occupied by the gas. However, when the vibration frequency is close to the natural frequency of the liquid inside the tank, measurement error increases, especially for high fill levels. The greatest effect from attitude disturbance on liquid mass gauging is through liquid sloshing. Therefore, liquid sloshing affects both the choice and location of the pressure transducers. Any liquid that flows into the transducer tubing increases the difficulty associated with processing the data.

3. Heating or cooling of the tank system influences the measurement accuracy. This is particularly obvious at low fill levels. The gauging accuracy does not change so much for normal propellants because of the small change in the thermodynamic properties. However, for cryogenic propellants or volatile liquids with the same heat load, the thermodynamic properties change drastically during the CMG operation time, which introduces large errors in the measurement. Modifications to the thermodynamic parameters should be considered by ensuring proper operating temperatures and pressures. The precise prediction of pressure rise and thermal stratification in a cryogenic tank needs to be studied for future cryogenic liquid mass gauging.

REFERENCES

[1] Beduz C, Rebiai R, Scurlock RG. Thermal Overfill and the surface vaporization of cryogenic liquids under storage conditions. Adv Cryog Eng 1984;29:795—804.

[2] Das SP, Chakraborty S, Dutta P. Studies on thermal stratification phenomena in LH_2 storage vessel. Heat Transf Eng 2004;25(4):54—66.

[3] Markham JR. High speed infrared radiation thermometer, system, and method. 2002. United States, http://www.osti.gov/scitech/servlets/purl/874313.

[4] Ishihara A, Sakai Y, Konishi K, Andoh E. Measurement of the burning surface temperature in Ammonium Perchlorate. Proc Combust Inst 2000;28:855—62.

[5] Suh NP, Tsai CL, Thompson Jr CL, Moore JS. Ignition and surface temperature of Double Base propellant at low pressure: I. Thermocouple measurements. AIAA J 1970;8(8):1314—21.

[6] Suh NP, Tsai CL. Thermocouple response characteristics in deflagrating low-conductivity materials. Trans ASME J Heat Transf 1971;93:77—87.

[7] Miller MS. An evaluation of Imbedded thermocouples as a solid-propellant combustion Diagnostics. In: Proceedings 21st JANNAF Combustion Meeting. CPIA Publication; 1989. p. 343—53.

[8] Ishihara A. An evaluation in temperature measurement of the burning surface by a thermocouple. AIAA; 2001. Paper 2001—3578.

[9] Seers P, Thomas W, Bruyere-Bergeron S. Determination of fuel droplet evaporation based on multiple thermocouple sizes. In: 49th AIAA Aerospace Sciences Meeting including the new Horizons Forum and Aerospace Exposition; 2011. AIAA 2011-789.

[10] Ishihara A. Temperature measurement of a burning surface by a thermocouple. J Propuls Power May–June 2004;20(3):455–9.

[11] Ishihara A. Temperature measurement of a burning surface by a thermocouple II. In: 41st AIAA/ASME/SAE/ASEE Joint Propulsion Conference & Exhibit; 2005. AIAA 2005-3576.

[12] Chato JC, A method for the measurement of thermal properties of biological materials, thermal problem in biotechnology, ASME Transactions, American Society of Mechanical Engineers, Winter annual meeting, December 3, 1968, New York, pp. 16–25.

[13] Balasubramaniam TA, Bowman HF. Thermal conductivity and thermal diffusivity of biomaterials: a simultaneous measurement technique. J Biomech Eng 1977;99 (3):148–54.

[14] Holeschovsky UB, Martin GT, Tester JW. A transient Spherical Source method to determine thermal conductivity of liquids and Gels. Int J Heat Mass Transf 1966;39(6):1135–40.

[15] Araki N, Futamura M, Makino A, Shibata H. Measurements of Thermophysical properties of Sodium Acetate hydrate. Int J Thermophys 1995;16(6):1455–66.

[16] Carslaw HS, Jaeger JC. Conduction of heat in solids. 2nd ed. Oxford, England, UK: Oxford University Press; 1959. p. 230–2.

[17] Ould Lahoucine C, Sakashita H, Kumada T. A method for measuring thermal conductivity of liquids and powders with a thermistor probe. Int Commun Heat Mass Transf 2003;30(4):445–54.

[18] Ould Lahoucine C, Sakashita H, Kumada T. Measurement of thermal conductivity of Buffer materials and evaluation of Existing correlations predicting it. Nucl Engn Des 2002;216(1–3):1–11.

[19] Ould Lahoucine C, Sakashita H, Kumada T. Simultaneous Determination of Thermophysical PropertiesUsing a thermistor, Part 1: Numerical model. AIAA J Thermophys Heat Transf 2004;18(3):295–301.

[20] Ould Lahoucine C, Sakashita H, Kumada T. Simultaneous Determination of Thermophysical properties using a thermistor, Part 2: experiment. J Thermophys Heat Transf 2004;18(3):302–8.

[21] Tegtmeier CL, Anusonti-Inthra P, Winchester JE. Analysis of a Turbine flow meter calibration Curve using CFD. In: 53rd AIAA Aerospace Sciences Meeting, AIAA 2015–1959; 2015.

[22] Morris AS, Langari R. Measurement and instrumentation. In: Theory and application, Chapter 16-flow measurement. 2nd ed. 2016. p. 493–529.

[23] Hansen EC, Restrepo JA. Development of a *small* vortex shedding flow meter for hypergolic propellants. 88-36-02 CP. In: 1st National Fluid Dynamics Conference, fluid dynamics and Co-located Conferences, OH, July; 1988. p. 1500–4.

[24] Cauty F, Comas P, Vuillot F. Magnetic flow meter measurement of solid propellant pressure-coupled responses using an acoustic analysis. J Propuls 1996;12(2):436–8.

[25] Matthijssen R, Van Put P. Ultrasonic flow meter for satellite propellant gauging and ground test facilities. In: 44th AIAA/ASME/SAE/ASEE Joint Propulsion Conference & Exhibit 21–23 July 2008. Hartford, CT: AIAA; 2008. 2008–4854.

[26] He D, Bai B, Zhang J, Wang X. Online measurement of gas and liquid flow rate in wet gas through one V-Cone throttle device. Exp Therm Fluid Sci 2016;75:129–36.

[27] Hepner DB, King CD, Littles JW. Zero gravity experiments in two-phase fluids flow regimes. In: ASME Intersociety Conf. Environmental Systems, San Francisco, CA. ASME; 1975. Paper No. 75-ENAs-24.

[28] Zhao L, Rezkallah KS. Gas–liquid flow patterns at microgravity. Int J Multiph Flow 1993;19:751–63.

[29] Kawahara A, Sadatomi M, Nei K, Matsuo H. Experimental Study on Bubble Velocity, void fraction and pressure drop for gas-liquid two-phase in a circular microchannel. Int J Heat Fluid Flow 2009;30:831−41.

[30] Luciani S, Brutin D, Le Niliot C, Rahli O, Tadrist L. Flow boiling in minichannels under normal, hyper-, and microgravity: local heat transfer analysis using inverse methods. J Heat Transf -Trans ASME 2008;130:1−13.

[31] Luciani S, Brutin D, Le Niliot C, Rahli O. Boiling heat transfer in a vertical micro-channel: local estimation during flow boiling with a non-intrusive method. Multiph Sci Technol 2009;21:297−328.

[32] Brutin D, Ajaev VS, Tadrist L. Pressure drop and void fraction during flow boiling in rectangular minichannels in weightlessness. Appl Therm Eng 2013;51:1317−27.

[33] Chen I, Downing R, Keshock EG, Al-Sharif M. Measurements and correlation of two-phase pressure drop under microgravity conditions. AIAA J Thermophys Heat Transf 1991;5:514−23.

[34] Choi B, Fujii T, Asano H, Sugimoto K. A study of gas−liquid two-phase flow in a horizontal tube under microgravity. Ann NY Acad Sci 2002;974:316−27.

[35] Zhao L, Rezkallah KS. Pressure drop in gas−liquid flow at microgravity conditions. Int J Multiph Flow 1995;21:837−49.

[36] Misawa M. An experimental and analytical investigation of flow boiling heat transfer under microgravity conditions. Ph.D. thesis. University of Florida; 1993.

[37] de Jong P, Gabriel KS. A preliminary study of two-phase annular flow at microgravity: experimental data of film thickness. Int J Multiph Flow 2003;29:1203−20.

[38] Godbole PV, Tang CC, Ghajar AJ. Comparison of void fraction correlations for different flow patterns in upward vertical two-phase flow. Heat Transf Eng 2011;32(10):843−60.

[39] Cioncolini A, Thome JR. Void fraction prediction in annular two- phase flow. Int J Multiph Flow 2012;43:72−84.

[40] Sakaguchi T, Shakutsui H, Mingawa H, Tomiyama A, Takahashi H. Pressure drop in gas-liquid-solid three-phase bubbly flow in vertical pipes, Advances in Multiphase flow. New York: Elsevier Science; 1995. p. 129−417.

[41] Sanaullah K, Zaidi HS, Hills JH. A study of bubbly flow using Resistivity Probes in A Novel configuration. Chem Engn J 2001;83(1):45−53.

[42] Sekoguchi H, Takeishi M, Kano H, Hironaga K, Nishimura T. Measurement of velocity and void fraction in gas-liquid two-phase flow with optical fiber. In: Proceedings 2nd International Symposium, application of laser anemometry to fluid mechanics; 1984. p. 1−5.

[43] Sekoguchi K, Takeishi M, Hironaga K, Nishimura T. Velocity measurement with electrical double-sensing devices in two-phase flow, measuring techniques in gas-liquid two-phase flows. Berlin: Springer; 1984.

[44] Song CH, Chung MK, No HC. Measurements of void fraction by an improved Multi-channel conductance void meter. Nucl Eng Des 1998;184(2):269−85.

[45] Kendoush AA, Sarkis ZA. Improving the accuracy of the capacitance method for void fraction measurement. Exp Therm Fluid Sci 1995;11(4):311−418.

[46] Hori K, Fujimoto T, Kawanishi K. Development of ultra-fast X-ray computed tomography scanner system. IEEE Trans Nucl Sci 1998;45(4):2089−94.

[47] Zhao Y, Bi Q, Yuan Y, Lv H. Void fraction measurement in steam−water two-phase flow using the gamma ray attenuation under high pressure and high temperature evaporating conditions. Flow Meas Instrum 2016;49:18−30.

[48] Mishima K, Hibiki T, Saito Y, Sugimoto J, Moriyama K. Visualization study of molten metal-water Interaction by using neutron radiography. Nucl Engn Des 1999;189 (1):391−403.

[49] Prasser HM, Boettger A, Zschau J. A new Electrode-mesh Tomograph for gas-liquid flows. Flow Meas Instrum 1998;9(2):111—9.
[50] Richter S, Aritomi M, Prasser HM, Hampel R. Approach towards spatial phase Reconstruction in transient bubbly flow using a wire-mesh sensor. Int J Heat Mass Transf 2002;45(5):1063—75.
[51] Rosensweig RE. Ferrohydrodynamics. Cambridge, U.K: Cambridge University Press; 1985. p. 131—76.
[52] Okubo M, Ishimoto J, Kamiyama S. Basic study on an energy Conversion system using gas-liquid two-phase flows of magnetic fluid, cavitation and multiphase flow. In: American Society of Mechanical Engineers, Vol. 194. Fairfield, NJ: FED; 1994. p. 105—10.
[53] Shuchi S, Mori T, Yamaguchi H. Flow boiling heat transfer of Binary mixed magnetic fluid. IEEE Trans Magn 2002;38(5):3234—6.
[54] Serizawa A, Feng Z, Kawara Z. Two-phase flow in microchannels. Exp Therm Fluid Sci 2002;26:703—14.
[55] Kawahara A, Chung PMY, Kawaji M. Investigation of two-phase flow pattern, void fraction and pressure drop in a microchannel. Int J Multiph Flow 2002;28:1411—35.
[56] Kawahara A, Sadatomi M, Okayama K, Kawaji M. Effects of liquid properties on pressure drop of two-phase gas-liquid flows through a microchannel. In: 1st International Conference on microchannels and minichannels, Rochester, New York, USA, April 24—25; 2003. p. 479—86. Paper No. ICMM2003-1058.
[57] Kawahara A, Sadatomi M, Okayama K, Kano K. Pressure drop for gas-liquid two-phase flow in microchannels-effects of channel size and liquid properties. In: Proceedings of the Third International Symposium on two-phase flow Modelling and Experimentation; 2004. 8 Pages in CD-ROM.
[58] Kawahara A, Sadatomi M, Okayama K, Kawaji M, Chung PMY. Effects of channel diameter and liquid properties on void fraction in adiabatic two-phase flow through microchannels. Heat Transf Eng 2005;26:13—9.
[59] Kawahara A, Sadatomi M, Okayama K, Kawaji M, Chung PMY. Assessment of void fraction correlations for adiabatic two-phase flows in microchannels. In: Proceedings of the Third International Conference on microchannels and minichannels; 2005. 8 Pages in CD-ROM.
[60] Kawahara A, Sadatomi M, Kumagae K. Effects of gas-liquid inlet/mixing conditions on two-phase flow in microchannels. Prog Multiph Flow Res 2006;1:197—203.
[61] Chung PMY, Kawaji M, Kawahara A, Shibata Y. Two-phase flow through square and circular microchannels -effects of channel geometry. Trans ASME, J Fluids Engn 2004;126:546—52.
[62] Chung PMY, Kawaji M. The effect of channel diameter on adiabatic two-phase flow characteristics in microchannels. Int J Multiph Flow 2004;Vol. 30:735—61.
[63] Kawaji M, Kawahara A, Mori K, Sadatomi M, Kumagae K. Gas-liquid two-phase flow in microchannels: the effects of gas-liquid injection methods. In: Proceedings of the 18th National and Seventh ISHMT-ASME Heat Transfer Conference; 2006. p. 80—9.
[64] Borowski SK. Nuclear thermal rocket and vehicle Options for lunar/Mars Transportation systems. In: Preprint, Conference on Advanced SEI Technologies, Cleveland, OH; September 1991.
[65] Corporation Bendix. Design development and manufacture of a breadboard radio frequency mass gauging system, Vol. 1; November 1974. Phase B Final Report., NASA CR-120620.
[66] Leaves KV. RF modal quantity gauging. N90-18483. UNCLAS; 1990. p. 471—8.
[67] Zimmerli GA. Radio frequency mass gauging of propellants. 2007. AIAA 2007-1198.
[68] Ambrose J, Yendler B, Collicott SH. Modeling to evaluate a spacecraft propellant gauging system. J Spacecr Rockets 2000;37(6):833—5.

[69] Boris Y, Steven HC, Timothy AM. Thermal gauging and rebalancing of propellant in multiple tank satellites. J Spacecr Rockets 2007;44:878—83.

[70] Purohit GP, Vu CC, Dhir VK. Transient lumped capacity thermodynamic model of satellite propellant tanks in micro-gravity. In: 37th Aerospace Sciences Meeting and Exhibit, Aerospace Sciences Meetings. AIAA; 1999. p. 99—1088.

[71] Narita T, Yendler B. Thermal propellant gauging system for BSS 601. AIAA; 2007. 2007—3149.

[72] Doux CJ, Justak JF. Liquid Oxygen test results for an optical mass gauge. AIAA; 2009. 2009—5393.

[73] Van Sciver SW, Adams T, Caimi F, Celik D, Justak J, Kocek D. Optical mass gauging of solid hydrogen. Cryogenics June—August 2004;44(6—8):501—6.

[74] Sullenberger RM, Munoz WM, Lyon MP, Vogel K, Yalin AP, Korman V, Polzin KA. Optical mass gauging system for measuring liquid levels in a reduced-gravity environment. J Spacecr Rockets 2011;48(3):528—33.

[75] Orazietti AJ, Orton GF. Propellant gauging for geostationary satellites. AIAA; 1986. 86—1716.

[76] Orton G. Low-G propellant gauge. N88-15937. UNCLAS; 1988. p. 253—66.

[77] Matthijssen R, van Put P. State-of-the Art, gauging components for improved propellant management on 3-AXIS stabilized spacecraft. AIAA; 2006. 2006—4714.

[78] Hansman RJJ. Fundamental limitations on low gravity fluid gauging technologies imposed by orbital mission requirement. AIAA; 1988. 88—3402.

[79] Trinks H. Assessment study of liquid content measurement methods applicable to space mission. 1984. Report TUHH-TRI-ESAA-84—2.

[80] Chapelon J, Shankar P, Newhouse V. Applications of the Double frequency technique in bubble sizing and pressure measurements in fluids, acoustical imaging. In: Proceedings of the 14[th]International Symposium; 1985. p. 753—5.

[81] Bupp FE. Development of a Zero-G gauging system, Vol. 1; December 1973. TRW Rept. 16740-6003-RU-OO (April Rept. TR-74-5).

[82] Mord AJ, Snyder HA, Kilpatrick KA, Hermanson LA, Hopkins RA, Vangundy DA. Fluid quantity gauging, Ball aerospace systems Rept. December 1988. DRD MA-183T, Contract NAS9—17616.

[83] Rogers AC, Dodge FT, Behering KA. Feasibility demonstration of a cryogenic fluid gauging for space vehicle applications. AIAA J Propuls Power 1994;11:980—5.

[84] Jurns JM, Rogers AC. Compression mass gauging in a liquid hydrogen dewar. 1995. NASA CR 198366.

[85] Dodge FT, Kuhl CA. LN$_2$ gauging test using the smart cryogenic CMG. Southwest Research Institute; 1998. Final Report, Project 04—8862, NASA-LeRC Contract C-1425-F.

[86] Green ST, Walter DB, Dodge FT. Ground testing of a compression mass gauge. In: 40th AIAA/ASME/SAE/ASEE Joint Propulsion Conference and Exhibit, 11—14 July 2004, Fort Lauderdale, Florida. AIAA; 2004. 2004—4151.

[87] Fu J, Chen XQ, Huang YY, Chen Y, Guo J. The simulation test of compression mass gauge for liquid propellant measurement. J Astronaut 2012;33(6):802—8.

[88] Monti R, Berry W. Liquid gauging in space: the G-22 Experiment. ESA J 1998;18:51—61.

[89] Abramson HN. The dynamic behavior of liquids in moving containers with applications to space vehicle technology. 1966. NASA SP-106.

[90] Veldman AEP, Vogels MES. Axisymmetric liquid sloshing under low-gravity conditions. Acta Astronaut 1984;10:641—9.

[91] Greensite AL. Analysis and design of space vehicle flight control systems. 1967. NASA CR-826.

[92] Barsi S. Ventless pressure control of cryogenics storage tanks. PhD thesis. Department of Mechanical and Aerospace Engineering, Case Western Reserve University; 2011.

Appendix 1: Governing Equations for Momentum, Mass, and Energy Transport

A1.1 CONTINUITY EQUATION (MASS CONSERVATION EQUATION)

The continuity equation (K.E.) expresses that mass is constant, or not destroyable. If one considers a mass balance for the volume element in Fig. A1.1, the following equation appears where ρ is the density of the fluid and u, v, and w are the velocity components in the x-, y-, and z-directions, respectively.

$$\frac{\partial \rho}{\partial \tau} + \frac{\partial(\rho u)}{\partial x} + \frac{\partial(\rho v)}{\partial y} + \frac{\partial(\rho w)}{\partial z} = 0 \qquad (A1.1)$$

Especially for steady-state $(\partial/\partial \tau \equiv 0)$ incompressible flow $(\rho = \text{constant})$ and two-dimensional $(w \equiv 0, \partial/\partial z \equiv 0)$ flow, one has

$$\frac{\partial u}{\partial x} + \frac{\partial v}{\partial y} = 0 \qquad (A1.2)$$

The derivation of Eqs. (A1.1) and (A1.2) can be found in, e.g., the book by Sunden [1].

According to Eq. (A1.1), the sum of the mass within the volume element dxdydz in Fig. A1.1 and the net in- or outflowing mass is constant. This mass can therefore be considered as a system from a thermodynamics point of view.

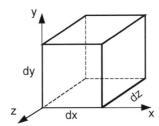

Figure A1.1 Volume element dxdydz.

A1.2 THE NAVIER–STOKES EQUATIONS

The equations of motion are derived from Newton's second law, which says that mass times acceleration in a certain direction is equal to the net external forces acting in the same direction. The external forces acting on a fluid element are split up into surface forces and volume or body forces. The volume or body forces act uniformly over the fluid element and are commonly generated by the gravity acceleration and electric or magnetic fields. The surface forces act as normal forces or shear forces on the boundary surface of the fluid element. Newton's second law can now be written as

$$\text{Mass} \times \text{Acceleration}_i = \text{Volume forces}_i + \text{Surface forces}_i \quad (A1.3)$$

The index i means direction i, i.e., x, y, or z.

The volume or body forces are calculated per unit mass and denoted by F_i.

$$F_i = (F_x, F_y, F_z)$$

The surface forces are calculated per unit area and are rather called stresses. A stress acting perpendicular to a surface (in the direction of the surface normal) is called a normal stress and that is acting along the surface is called a shear stress. In Fig. A1.2 the stresses on a two-dimensional element dxdy are shown. σ_{xx} and σ_{yy} are normal stresses in the x- and y-directions, respectively. The shear stresses are denoted as σ_{xy} and σ_{yx}, where the first index indicates the axis to which the considered surface is directed perpendicularly and the second index gives the direction of the stress.

The schematic in Fig. A1.2 can easily be developed to the three-dimensional case but then the normal stress σ_{zz} and the shear stresses σ_{xz}, σ_{zx} and σ_{yz}, σ_{zy} have to be added.

If the various terms in Eq. (A1.3) are set up for the x-, y-, and z-directions, one finds [1–4]

$$\hat{x}: \ \rho\left(\frac{\partial u}{\partial \tau} + u\frac{\partial u}{\partial x} + v\frac{\partial u}{\partial y} + w\frac{\partial u}{\partial z}\right) = \rho F_x + \frac{\partial \sigma_{xx}}{\partial x} + \frac{\partial \sigma_{yx}}{\partial y} + \frac{\partial \sigma_{zx}}{\partial z} \quad (A1.4)$$

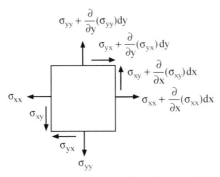

Figure A1.2 Stresses on the element dxdy.

$$\widehat{y}: \; \rho\left(\frac{\partial v}{\partial \tau} + u\frac{\partial v}{\partial x} + v\frac{\partial v}{\partial y} + w\frac{\partial v}{\partial z}\right) = \rho F_y + \frac{\partial \sigma_{xy}}{\partial x} + \frac{\partial \sigma_{yy}}{\partial y} + \frac{\partial \sigma_{zy}}{\partial z} \quad (A1.5)$$

$$\widehat{z}: \; \rho\left(\frac{\partial w}{\partial \tau} + u\frac{\partial w}{\partial x} + v\frac{\partial w}{\partial y} + w\frac{\partial w}{\partial z}\right) = \rho F_z + \frac{\partial \sigma_{xz}}{\partial x} + \frac{\partial \sigma_{yz}}{\partial y} + \frac{\partial \sigma_{zz}}{\partial z} \quad (A1.6)$$

A1.2.1 The Stress Tensor σ_{ij}

The nine stresses σ_{xx}, σ_{xy}, ... , σ_{zz} can be brought together in the tensor σ_{ij} (matrix with three rows and three columns) as

$$\sigma_{ij} = \begin{bmatrix} \sigma_{xx} & \sigma_{xy} & \sigma_{xz} \\ \sigma_{yx} & \sigma_{yy} & \sigma_{yz} \\ \sigma_{zx} & \sigma_{zy} & \sigma_{zz} \end{bmatrix} \quad (A1.7)$$

The mean value of the normal stresses is defined as the fluid pressure.

$$-p = \frac{\sigma_{xx} + \sigma_{yy} + \sigma_{zz}}{3} = \frac{\sigma_{ii}}{3} \quad (A1.8)$$

where the negative sign is necessary because of sign convention. This definition of the pressure is the same as the one being used for fluids at rest. For the case with a moving fluid, the definition is arbitrary but convenient. The quantity p according to Eq. (A1.8) is however not depending on the chosen coordinate system but has the same value in a Cartesian x-, y-, and z-coordinate system as in a cylindrical coordinate system r, θ, z. A quantity with this property is called an invariant.

The indices i and j in σ_{ij} can be x, y, or z. If an index is repeated as in σ_{ii} (Eq. (A1.8)), this means that summation over the repeated index should be carried out, i.e., $\sigma_{ii} = \sigma_{xx} + \sigma_{yy} + \sigma_{zz}$.

The stress tensor σ_{ij} is usually written as

$$\sigma_{ij} = -p\delta_{ij} + d_{ij} \tag{A1.9}$$

where

d_{ij} is called the deviatoric stress tensor

δ_{ij} is the Kronecker delta $\begin{cases} 0 & i \neq j \\ 1 & i = j \end{cases}$.

From the definitions in Eqs. (A1.8) and (A1.9), it follows (note $\delta_{ii} = 3$)

$$d_{ii} = d_{xx} + d_{yy} + d_{zz} = 0 \tag{A1.10}$$

Especially for a Newtonian fluid, one writes

$$d_{ij} = 2\mu(e_{ij} - \Delta\delta_{ij}/3) \tag{A1.11}$$

where μ is the dynamic viscosity [kg/(m s)] and

$$e_{ij} = \text{strain} - \text{rate} - \text{tensor} = \frac{1}{2}\left(\frac{\partial u_i}{\partial x_j} + \frac{\partial u_j}{\partial x_i}\right) \tag{A1.12}$$

$$\Delta = e_{ii} = \frac{\partial u_i}{\partial x_i} = \frac{\partial u}{\partial x} + \frac{\partial v}{\partial y} + \frac{\partial w}{\partial z} \tag{A1.13}$$

The derivation of these equations can be found in, e.g., Batchelor [3] and Young et al. [4]. Originally the Eq. (A1.11) comes from Navier (1822), Poisson (1829), Saint-Venant (1843), and Stokes (1845).

For water and most gases the assumption of a Newtonian fluid corresponding to Eq. (A1.11) is a valid approach.

It should be noted that the pressure p is defined as $-p = \sigma_{ii}/3$ and that it is not self-evident that this pressure is equal to the one in thermodynamic relations. However, if the flow is incompressible, the defined pressure is equal to the thermodynamic pressure, see Batchelor [3] and Kestin [5].

If Eqs. (A1.9), (A1.11), and (A1.12) are substituted in Eqs. (A1.4)−(A1.6), one finds the Navier−Stokes equations (N.S.).

A1.2.2 The Navier–Stokes Equations for Two-Dimensional and Incompressible Flows

If the flow is incompressible one has $\rho = $ constant and $\Delta = 0$.

For the stresses one finds by using Eqs. (A1.9), (A1.11), and (A1.12),

$$\sigma_{xx} = -p + 2\mu\, e_{xx} = -p + 2\mu\frac{\partial u}{\partial x} \tag{A1.14}$$

$$\sigma_{xy} = \sigma_{yx} = 2\mu\, e_{xy} = \mu\left(\frac{\partial u}{\partial y} + \frac{\partial v}{\partial x}\right) \tag{A1.15}$$

$$\sigma_{yy} = -p + 2\mu\, e_{yy} = -p + 2\mu\frac{\partial v}{\partial y} \tag{A1.16}$$

By substituting Eq. (A1.2), (A1.14) and (A1.15) in Eq. (A1.4), and Eqs. (A1.2), (A1.15) and (A1.16) in Eq. (A1.5), one finds

$$\widehat{x}: \ \rho\left(\frac{\partial u}{\partial \tau} + u\frac{\partial u}{\partial x} + v\frac{\partial u}{\partial y}\right) = \rho F_x - \frac{\partial p}{\partial x} + \mu\left(\frac{\partial^2 u}{\partial x^2} + \frac{\partial^2 u}{\partial y^2}\right) \tag{A1.17}$$

$$\widehat{y}: \ \rho\left(\frac{\partial v}{\partial \tau} + u\frac{\partial v}{\partial x} + v\frac{\partial v}{\partial y}\right) = \rho F_y - \frac{\partial p}{\partial x} + \mu\left(\frac{\partial^2 v}{\partial x^2} + \frac{\partial^2 v}{\partial y^2}\right) \tag{A1.18}$$

In Eqs. (A1.17) and (A1.18) the dynamic viscosity μ has been assumed constant.

A1.2.3 Derivation of the Complete Temperature Field Equation

Consider the momentum and energy exchange for the volume element dxdydz in Fig. A1.3. According to the mass conservation Eq. (A1.1), the

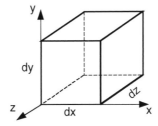

Figure A1.3 Volume element dxdydz.

mass inside the element and the incoming and outgoing masses form a thermodynamic system (total mass constant). For this mass the first law of thermodynamics is applicable, i.e.,

$$\Delta \dot{E} = \dot{Q} - \dot{W} \qquad (A1.19)$$

A1.2.3.1 Determination of $\Delta \dot{E}$

The total energy change $\Delta \dot{E}$ can be written as

$$\Delta \dot{E} = \Delta \dot{E}_{in-out} + \Delta \dot{E}_{dxdydz} \qquad (A1.20)$$

For $\Delta \dot{E}_{dxdydz}$, one has

$$\Delta \dot{E}_{dxdydz} = \frac{\partial}{\partial \tau}(\rho e^{0})dxdydz \qquad (A1.21)$$

where e^0 is the energy per mass unit.

$\Delta \dot{E}_{in-out}$ can be written as

$$\Delta \dot{E}_{in-out} = \frac{\partial}{\partial x}(\rho u e^{0})dxdydz + \frac{\partial}{\partial y}(\rho v e^{0})dxdydz + \frac{\partial}{\partial z}(\rho w e^{0})dxdydz$$

$$(A1.22)$$

where the first term expresses the net transport in the x-direction and the second and third terms are the net transports in the y- and z-directions, respectively.

The energy e^0 is the sum of the internal energy e and the kinetic energy (per mass unit) $V^2/2 = (u^2 + v^2 + w^2)/2$.

Thus one finds

$$e^{0} = e + \frac{V^2}{2} \qquad (A1.23)$$

The energy change $\Delta \dot{E}$ can now be expressed as

$$\Delta \dot{E} = \left[\frac{\partial}{\partial \tau}(\rho(e + V^2/2)) + \frac{\partial}{\partial x}(\rho u(e + V^2/2)) + \frac{\partial}{\partial y}(\rho v(e + V^2/2))\right.$$

$$\left. + \frac{\partial}{\partial z}(\rho w(e + V^2/2))\right]dxdydz$$

$$(A1.24)$$

A1.2.3.2 Determination of the Heat Transfer Rate $\dot{Q}$

The heat transfer rate $\dot{Q}$ is the rate of heat conduction in the fluid and can be found by a similar procedure as in Ref. [1], i.e.,

$$\dot{Q} = \left(\Delta\dot{Q}_x + \Delta\dot{Q}_y + \Delta\dot{Q}_z \right)$$
$$= \left(\frac{\partial}{\partial x} \left(k \frac{\partial T}{\partial x} \right) + \frac{\partial}{\partial y} \left(k \frac{\partial T}{\partial y} \right) + \frac{\partial}{\partial z} \left(k \frac{\partial T}{\partial z} \right) \right) \tag{A1.25}$$

A1.2.3.3 Determination of the Work Rate $\dot{W}$

Consider at first the element dxdy in the xy-plane according to Fig. A1.4.

The stresses σ_{xx}, σ_{yx}, and σ_{zx} act in the x-direction. These forces perform a work on the fluid within the element. For the normal stress σ_{xx}, one has, as indicated in Fig. A1.4,

$$\rightarrow \left(\sigma_{xx}u + \frac{\partial}{\partial x}(\sigma_{xx}u)dx \right) dydz$$

$$\leftarrow \sigma_{xx}u \, dydz$$

The net contribution from σ_{xx} is $\dfrac{\partial}{\partial x}(\sigma_{xx}u)dxdydz$

Similarly the shear stresses σ_{yx} and σ_{zx} in the x-direction perform works according to

$$\text{Net contribution from } \sigma_{yx}: \frac{\partial}{\partial y}(\sigma_{yx}u)dxdydz$$

$$\text{Net contribution from } \sigma_{zx}: \frac{\partial}{\partial z}(\sigma_{zx}u)dxdydz$$

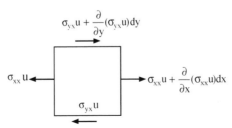

Figure A1.4 The element dxdy in the xy-plane work by forces in the x-direction.

In Eq. (A1.19), $\dot{W}$ is positive if the system performs work. In this case the surrounding is performing work on the system and thus the work is negative. One then has

$$\dot{W}_x = -\left(\frac{\partial}{\partial x}(\sigma_{xx}u) + \frac{\partial}{\partial y}(\sigma_{yx}u) + \frac{\partial}{\partial z}(\sigma_{zx}u)\right)dxdydz \qquad (A1.26)$$

$$\dot{W}_y = -\left(\frac{\partial}{\partial x}(\sigma_{xy}v) + \frac{\partial}{\partial y}(\sigma_{yy}v) + \frac{\partial}{\partial z}(\sigma_{zy}v)\right)dxdydz \qquad (A1.27)$$

$$\dot{W}_z = -\left(\frac{\partial}{\partial x}(\sigma_{xz}w) + \frac{\partial}{\partial y}(\sigma_{yz}w) + \frac{\partial}{\partial z}(\sigma_{zz}w)\right)dxdydz \qquad (A1.28)$$

The work by the surface forces becomes

$$\dot{W}_{surface} = \dot{W}_x + \dot{W}_y + \dot{W}_z$$

If tensorial formulation is used, this work can be written as

$$\dot{W}_{surface} = -\frac{\partial}{\partial x_j}(\sigma_{ji}u_i)dxdydz \qquad (A1.29a)$$

In addition, if a volume force $F_i = (F_x, F_y, F_z)$ is acting on the volume element, its contribution to the work can be written as

$$\dot{W}_{volume} = -(F_x u + F_y v + F_z w)\rho \, dxdydz = -F_i u_i \rho \, dxdydz \qquad (A1.29b)$$

The total work is governed by

$$\dot{W} = \dot{W}_{surface} + \dot{W}_{volume}$$

A1.2.3.3.1 The Energy Equation in its Primary Form

Eqs. (A1.24), (A1.25), and (A1.29a) and (A1.29b) are now inserted in Eq. (A1.19). If the tensorial notation is used,

$$\frac{\partial}{\partial \tau}\left(\rho\left(e + \frac{V^2}{2}\right)\right) + \frac{\partial}{\partial x_i}\left(\rho u_i\left(e + \frac{V^2}{2}\right)\right)$$
$$= \frac{\partial}{\partial x_i}\left(k\frac{\partial T}{\partial x_i}\right) + \frac{\partial}{\partial x_j}(\sigma_{ji}u_i) + \rho F_i u_i \qquad (A1.30)$$

Here Eq. (A1.30) is called the primary form of the energy equation.

A1.2.3.3.2 Rewriting the Energy Equation

Eq. (A1.30) will now be transferred to a more appropriate form. The left hand side of Eq. (A1.30) can be written as

$$\frac{\partial}{\partial \tau}(\rho e) + \frac{\partial}{\partial x_i}(\rho u_i e) + \frac{\partial}{\partial \tau}\left(\rho \frac{V^2}{2}\right) + \frac{\partial}{\partial x_i}\left(\rho u_i \frac{V^2}{2}\right) \qquad (A1.31)$$

By applying the mass conservation Eq. (A1.1), i.e.,

$$\frac{\partial \rho}{\partial \tau} + \frac{\partial}{\partial x_i}(\rho u_i) = 0$$

Eq. (A1.31) can be written as

$$\rho \frac{\partial e}{\partial \tau} + \rho u_i \frac{\partial e}{\partial x_i} + \rho \frac{\partial}{\partial \tau}\left(\frac{V^2}{2}\right) + \rho u_i \frac{\partial}{\partial x_i}\left(\frac{V^2}{2}\right) \qquad (A1.32)$$

The Navier—Stokes Eqs. (A1.4)—(A1.6), can be written by tensorial notation as

$$\rho \frac{\partial u_i}{\partial \tau} + \rho u_j \frac{\partial u_i}{\partial x_j} = \rho F_i + \frac{\partial}{\partial x_j}(\sigma_{ji}) \qquad (A1.33)$$

If Eq. (A1.33) is multiplied by u_i, one finds

$$\rho \frac{\partial}{\partial \tau}\left(\frac{u_i u_i}{2}\right) + \rho u_j \frac{\partial}{\partial x_j}\left(\frac{u_i u_i}{2}\right) = \rho F_i u_i + u_i \frac{\partial}{\partial x_j}(\sigma_{ji})$$

However, $u_i u_i/2 = (u^2 + v^2 + w^2)/2$, and hence it is possible to write the above equation as

$$\rho \frac{\partial}{\partial \tau}\left(\frac{V^2}{2}\right) + \rho u_j \frac{\partial}{\partial x_j}\left(\frac{V^2}{2}\right) = \rho F_i u_i + u_i \frac{\partial}{\partial x_j}(\sigma_{ji}) \qquad (A1.34)$$

Substituting Eq. (A1.34) in Eq. (A1.32) gives

$$\rho \frac{\partial e}{\partial \tau} + \rho u_i \frac{\partial e}{\partial x_i} + \rho F_i u_i + u_i \frac{\partial}{\partial x_j}(\sigma_{ji}) \qquad (A1.35)$$

If Eq. (A1.35) is introduced into Eq. (A1.30), the energy equation can be written as

$$\rho \frac{\partial e}{\partial \tau} + \rho u_i \frac{\partial e}{\partial x_i} = \frac{\partial}{\partial x_i}\left(k \frac{\partial T}{\partial x_i}\right) + \sigma_{ji} \frac{\partial u_i}{\partial x_j} \qquad (A1.36)$$

Eq. (A1.36) is a general form of the energy equation. The last term on the right hand side of Eq. (A1.36) will now be rewritten. The stress tensor σ_{ji} can be written as

$$\sigma_{ji} = -p\delta_{ji} + \mu\left(\frac{\partial u_j}{\partial x_i} + \frac{\partial u_i}{\partial x_j} - \frac{2}{3}\Delta\delta_{ij}\right) \qquad (A1.37)$$

The term $\sigma_{ji}\frac{\partial u_i}{\partial x_j}$ can now be written as

$$\sigma_{ji}\frac{\partial u_i}{\partial x_j} = -p\frac{\partial u_i}{\partial x_i} + \mu\frac{\partial u_i}{\partial x_j}\left(\frac{\partial u_j}{\partial x_i} + \frac{\partial u_i}{\partial x_j}\right) - \frac{2}{3}\mu\Delta\frac{\partial u_i}{\partial x_i} \qquad (A1.38)$$

The rate of expansion Δ can be written as $\Delta = \frac{\partial u_i}{\partial x_i}$. If Eq. (A1.38) is inserted in Eq. (A1.36), one finds

$$\rho\frac{\partial e}{\partial \tau} + \rho u_i\frac{\partial e}{\partial x_i} = \frac{\partial}{\partial x_i}\left(k\frac{\partial T}{\partial x_i}\right) - p\Delta + \mu\frac{\partial u_i}{\partial x_j}\left(\frac{\partial u_j}{\partial x_i} + \frac{\partial u_i}{\partial x_j}\right) - \frac{2}{3}\mu\Delta^2 \qquad (A1.39)$$

Eq. (A1.39) is one of the many possible forms of the energy equation. However, additional rearrangement is necessary to have it more handy. The internal energy e can be expressed by introducing the enthalpy h according to the equation $e = h - p/\rho$. The left hand side of Eq. (A1.39) can therefore be written as

$$\rho\frac{\partial e}{\partial \tau} + \rho u_i\frac{\partial e}{\partial x_i} = \rho\frac{\partial h}{\partial t} + \rho u_i\frac{\partial h}{\partial x_i} - \frac{\partial p}{\partial \tau} + \frac{p}{\rho}\frac{\partial \rho}{\partial \tau} - u_i\frac{\partial p}{\partial x_i} + \frac{p}{\rho}u_i\frac{\partial \rho}{\partial x_i}$$

$$= \rho\left(\frac{\partial h}{\partial \tau} + u_i\frac{\partial h}{\partial x_i}\right) - \frac{\partial p}{\partial \tau} - u_i\frac{\partial p}{\partial x_i} + \frac{p}{\rho}\left(\frac{\partial \rho}{\partial \tau} + u_i\frac{\partial \rho}{\partial x_i}\right) \qquad (A1.40)$$

By considering the mass conservation equation, the last term in Eq. (A1.40) can be written as

$$\frac{\partial \rho}{\partial \tau} + u_i\frac{\partial \rho}{\partial x_i} = -\rho\frac{\partial u_i}{\partial x_i} = -\rho\Delta$$

Finally, one then has

$$\rho\frac{\partial e}{\partial \tau} + \rho u_i\frac{\partial e}{\partial x_i} = \rho\left(\frac{\partial h}{\partial \tau} + u_i\frac{\partial h}{\partial x_i}\right) - \left(\frac{\partial p}{\partial \tau} + u_i\frac{\partial p}{\partial x_i}\right) - p\Delta \qquad (A1.41)$$

Substituting Eq. (A1.41) in Eq. (A1.39) gives

$$\rho\left(\frac{\partial h}{\partial \tau} + u_i\frac{\partial h}{\partial x_i}\right) = \frac{\partial}{\partial x_i}\left(k\frac{\partial T}{\partial x_i}\right) + \frac{\partial p}{\partial \tau} + u_i\frac{\partial p}{\partial x_i} + \mu\frac{\partial u_i}{\partial x_j}\left(\frac{\partial u_j}{\partial x_i} + \frac{\partial u_i}{\partial x_j}\right) - \frac{2}{3}\mu\Delta^2$$

(A1.42)

Eq. (A1.42) is the general form of the energy equation suitable for continued simplifications.

If the enthalpy $h = c_p T$ is introduced, the so-called general form of the temperature field equation is obtained, i.e.,

$$\rho c_p\left(\frac{\partial T}{\partial \tau} + u_i\frac{\partial T}{\partial x_i}\right) = \frac{\partial}{\partial x_i}\left(k\frac{\partial T}{\partial x_i}\right) + \frac{\partial p}{\partial \tau} + u_i\frac{\partial p}{\partial x_i} + \mu\frac{\partial u_i}{\partial x_j}\left(\frac{\partial u_j}{\partial x_i} + \frac{\partial u_i}{\partial x_j}\right)$$
$$- \frac{2}{3}\mu\Delta^2$$

(A1.43)

A1.3 THE BOUNDARY LAYER FORM OF THE TEMPERATURE FIELD EQUATION

The temperature field equation for a two-dimensional boundary layer is given without a derivation. This equation is written as

$$\rho c_p\left(\frac{\partial T}{\partial \tau} + u\frac{\partial T}{\partial x} + v\frac{\partial T}{\partial y}\right) = \frac{\partial}{\partial y}\left(k\frac{\partial T}{\partial y}\right) + \frac{\partial p}{\partial \tau} + u\frac{\partial p}{\partial x} + \mu\left(\frac{\partial u}{\partial y}\right)^2 \quad \text{(A1.44)}$$

The last three terms in the right hand side are commonly neglected but the last one of these is the most important particularly at high flow velocities (frictional heating).

The final form is then given by

$$\rho c_p\left(\frac{\partial T}{\partial \tau} + u\frac{\partial T}{\partial x} + v\frac{\partial T}{\partial y}\right) = \frac{\partial}{\partial y}\left(k\frac{\partial T}{\partial y}\right) + \mu\left(\frac{\partial u}{\partial y}\right)^2 \quad \text{(A1.45)}$$

A1.4 BOUNDARY LAYER EQUATIONS FOR THE LAMINAR CASE

The mathematic difficulties in solving the flow field Eqs. (A1.1) and (A1.4)–(A1.6), and the temperature field Eq. (A1.43), have forced researchers and engineers to develop ideas to simplify the equations. The so-called boundary layer theory according to Prandtl [6] was found to be successful and has then been applied for many engineering problems. In the boundary layer theory, the flow field is divided into two regions: (1) a thin layer adjacent to the surface of the body where the friction forces are strong and high gradients in the flow and temperature fields exist and (2) the region outside the boundary layer where the friction forces are negligible. The latter region is called the potential flow field or the outer region. The gradients in the flow and temperature fields are relatively small in this region.

A basic assumption in the boundary layer theory is that the fluid immediately at the surface is at rest in relation to the body. This assumption is valid except at very low pressures, as then the mean free path of the molecular motion is of the same order of magnitude as the overall dimension of the body. Thus the boundary layer can be defined as the region where the flow velocity is changed from the potential flow value to zero at the body surface (Fig. A1.5).

Similarly the thermal boundary layer or the temperature boundary layer can be defined as the region where the fluid temperature is changed from the value in the potential flow to that of the surface (Fig. A1.6).

If the boundary layer thicknesses δ and δ_T are very small compared to other dimensions, one has for a two-dimensional boundary layer [7]

$$u \gg v \tag{A1.46}$$

$$\frac{\partial u}{\partial y} \gg \frac{\partial u}{\partial x}, \frac{\partial v}{\partial x}, \frac{\partial v}{\partial y} \tag{A1.47}$$

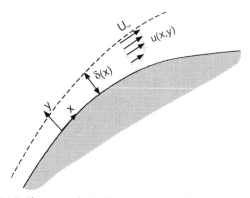

Figure A1.5 Flow or velocity boundary layer along a body surface.

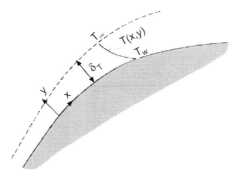

Figure A1.6 The thermal boundary layer or temperature boundary layer along a body surface.

$$\frac{\partial T}{\partial y} >> \frac{\partial T}{\partial x} \qquad (A1.48)$$

From the Navier–Stokes equation in the y-direction (Eq. (A1.18)), it is found that the pressure p is independent of y, i.e.,

$$p = p(x) \qquad (A1.49)$$

The Navier–Stokes equation in the x-direction is simplified to

$$\rho\left(u\frac{\partial u}{\partial x} + v\frac{\partial u}{\partial y}\right) = \rho F_x - \frac{dp}{dx} + \mu\frac{\partial^2 u}{\partial y^2} \qquad (A1.50)$$

The temperature field Eq. (A1.45) (T.E.) is simplified to

$$u\frac{\partial T}{\partial x} + v\frac{\partial T}{\partial y} = \frac{k}{\rho c_p}\frac{\partial^2 T}{\partial y^2} \qquad (A1.51)$$

For the inviscid flow, the potential flow, outside the boundary layer the Bernoulli equation is valid, i.e.,

$$p + \frac{1}{2}\rho U^2 = \text{Constant} \qquad (A1.52)$$

where U is the potential flow velocity. With Eq. (A1.52), the pressure term in Eq. (A1.50) can be written as

$$\frac{dp}{dx} = -\rho U\frac{dU}{dx} \qquad (A1.53)$$

Commonly the Prandtl number, Pr, is introduced as

$$\text{Pr} = \frac{\nu \rho c_p}{k} = \frac{\mu c_p}{k} \quad (A1.54)$$

In Eq. (A1.54), ν is the kinematic viscosity (m^2/s) defined as $\nu = \mu/\rho$.

In summary the following equations are valid for a two-dimensional laminar boundary layer at steady flow conditions and by neglecting the mass or body forces:

$$\text{K.E.} \quad \frac{\partial u}{\partial x} + \frac{\partial v}{\partial y} = 0 \quad (A1.55)$$

$$\text{N.S.} \quad u\frac{\partial u}{\partial x} + v\frac{\partial u}{\partial y} = U\frac{dU}{dx} + \frac{\mu}{\rho}\frac{\partial^2 u}{\partial y^2} \quad (A1.56)$$

$$\text{T.E.} \quad u\frac{\partial T}{\partial x} + v\frac{\partial T}{\partial y} = \frac{\mu}{\rho \text{Pr}}\frac{\partial^2 T}{\partial y^2} \quad (A1.57)$$

Eqs. (A1.55)−(A1.57) are only valid for laminar cases because the flow and temperature fields are assumed to be independent of time. If the curvature of the body surface is modest and if x is along the surface and y is normal to the surface, the equations will be valid [7].

The boundary layer can be either laminar or turbulent depending on the flow conditions. Fig. A1.7 illustrates these two types of the velocity boundary layer along a plane plate. At the leading edge a laminar boundary layer develops. At a certain distance, say x_c, disturbances appear and these grow within the boundary layer and then transition from the laminar to turbulent boundary layer flow takes place. The distance beyond which the flow field is no longer laminar is determined by the so-called critical Reynolds number, $\text{Re}_c = U_\infty x_c/\nu$. A typical value is $\text{Re}_c = 5 \cdot 10^5$.

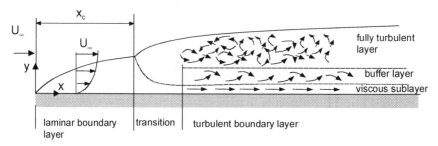

Figure A1.7 Laminar and turbulent boundary layers over a plane plate.

If the plate surface is rough, the transition may occur at a much lower value of Reynolds number, e.g., 10^5. In other cases, the plate surface might be extremely smooth, and for ideal situations, the transition might be delayed and the laminar boundary layer can stay up to $Re_c = 5 \cdot 10^6$.

A1.5 DIMENSIONLESS GROUPS AND RULES OF SIMILARITY

As stated previously, it is extremely difficult to solve the governing equations for convective flow and heat transfer. Experimental investigations have dominated in the past and results have been presented in terms of empirical correlations, which include dimensionless numbers. The advantage of using dimensionless groups is that many variables can be combined with a limited number of dimensionless numbers. Thus it is important to find the dimensionless groups that are relevant to a certain heat transfer problem. Two different methods are commonly used. In one method, all variables that affect the heat transfer process are set up and then the number of independent dimensionless groups is determined by the so-called Buckingham Π theorem [8]. The procedure is simple but the analysis can result in errors if any of the variables are missing. In another method, the dimensionless groups are determined from the dimensionless form of the governing equations. In this chapter a principle description based on the second method is provided.

In the following, nondimensional variables are introduced.

$$\overline{x} = \frac{x}{L}, \overline{y} = \frac{y}{L}, \overline{z} = \frac{z}{L}$$

$$\overline{u} = \frac{u}{U_0}, \overline{v} = \frac{v}{U_0}, \overline{w} = \frac{w}{U_0} \tag{A1.58}$$

$$\overline{p} = \frac{p}{\rho U_0^2}, \overline{\tau} = \frac{\tau U_0}{L}, \overline{T} = \frac{T - T_w}{T_\infty - T_w}$$

where U_0 is a characteristic velocity and L is a characteristic dimension (length). It is possible to show that the solutions to the flow and temperature fields have the forms

$$\overline{u} = f_1(\overline{x}, \overline{y}, \overline{z}, \overline{\tau}, Re) \tag{A1.59}$$

$$\overline{v} = f_2(\overline{x}, \overline{y}, \overline{z}, \overline{\tau}, Re) \tag{A1.60}$$

$$\overline{w} = f_3(\overline{x}, \overline{y}, \overline{z}, \overline{\tau}, Re) \tag{A1.61}$$

$$\overline{p} = f_4(\overline{x}, \overline{y}, \overline{z}, \overline{\tau}, Re) \tag{A1.62}$$

$$\overline{T} = f_5(\overline{x}, \overline{y}, \overline{z}, \overline{\tau}, Re, Pr) \tag{A1.63}$$

where

$$Re = \frac{U_0 L}{\nu} \tag{A1.64}$$

Eqs. (A1.59)–(A1.63) are valid if the mass or body force is excluded, which means that only forced convection is considered.

From Eqs. (A1.59)–(A1.62), it is possible to find that the drag coefficient c_D and the lift coefficient c_L only depend on the Reynolds number, Re, see, e.g., [2].

For the heat transfer coefficient α or h, by using Eqs. (A1.58) and (A1.63), one obtains

$$h = \frac{1}{L} k_f \left(\frac{\partial \overline{T}}{\partial \overline{y}} \right)_{y=0}$$

or

$$\frac{hL}{k_f} = \left(\frac{\partial \overline{T}}{\partial \overline{y}} \right)_{y=0} \tag{A1.65}$$

The group hL/k_f is called the Nusselt number, Nu. By applying Eqs. (A1.59)–(A1.63), one can write

$$Nu = \frac{hL}{k_f} = f_6(\overline{x}, \overline{y}, \overline{z}, \overline{\tau}, Re, Pr) \tag{A1.66}$$

If steady flow conditions are considered, a mean value of the heat transfer coefficient can be written as

$$Nu = f_7(Re, Pr) \tag{A1.67}$$

The so-called rules of similarity can be formulated as follows:
- The flow field around or inside geometrically similar bodies is identical if the Reynolds number is equal and if dimensionless variables are applied (the boundary conditions need to be the same).
- The temperature field around or inside geometrically similar bodies is identical if the Reynolds number and the Prandtl number are equal and if dimensionless variables are applied (the boundary conditions need to be the same).

A general dimensional analysis for a fluid of variable properties is not possible. However, here the result of a simplified analysis for a gas is presented. The thermophysical properties are assumed as follows:

$$\rho = \frac{p}{RT}, \mu = C_\mu T^a, k = C_k T^a, c_p = \text{Constant}, \text{Pr} = \text{Constant} \quad \text{(A1.68)}$$

By a thorough analysis of the governing equations [9], it is possible to find that the average Nusselt number value should follow an equation such as

$$\overline{\text{Nu}} = f\left(\text{Re}, \text{Pr}, \text{Ma}, \frac{T_w}{T_0}, a, \gamma \right) \quad \text{(A1.69)}$$

In this case, six parameters appear in the function f and accordingly the number of situations where physical similarity can be achieved is reduced.

The Eckert number (Ec) $\left(\frac{U^2}{c_p(T_w - T_0)} \right)$ is an important parameter in high-speed heat transfer but it can be shown to be related to the Mach number, the specific heat ratio γ, and the ratio of the surface temperature to the reference temperatures. Then it does not appear directly in Eq. (A1.69).

The more accurate the thermophysical properties are, the more constants will be added to Eq. (A1.69). Then even more severe restrictions to achieve similarity appear.

For further reading, Refs. [10—12] are recommended.

REFERENCES

[1] Sunden B. Introduction to heat transfer. UK: WIT Press; 2012.
[2] White FM. Fluid mechanics. 8th ed. New York: McGraw-Hill; 2016.
[3] Batchelor GK. An introduction to fluid dynamics. 2nd ed. Cambridge: Cambridge University Press; 1970.
[4] Young DF, Munson BR, Okiishi TH. A brief introduction to fluid mechanics. 5th ed. J. Wiley & Sons; 2011.
[5] Kestin J. A course in thermodynamics, vol. II. Blaisdell Publ. Co; 1966. Waltham, Massachusetts.
[6] Prandtl L. Über Flüssigkeitsbewegung bei sehr kleiner Reibung. In: Proc. Third Int. Math. Congr; 1904. p. 484—91. Heidelberg.
[7] Schlichting H. Boundary layer theory. 7th ed. New York: McGraw-Hill; 1979.
[8] W Bluman G, Cole JD. Similarity methods for differential equations. Applied mathematical sciences, vol. 13. New York: Springer-Verlag; 1974.
[9] Eckert ERG, Drake Jr RM. Analysis of heat and mass transfer. New York: McGraw-Hill; 1972.
[10] Cengel YA, Boles MA. Thermodynamics: an engineering approach. 7th ed. New York: McGraw-Hill; 2011.
[11] Kestin J. A course in thermodynamics, vol. I. Blaisdell Publ. Co; 1966. Waitham, Massachusetts.
[12] Hughes WF, Gaylord EW. Basic equations of engineering science, Schaum's outline series. New York: McCraw-Hill; 1964.

Appendix 2: Dimensionless Numbers of Relevance in Aerospace Heat Transfer

Biot number, $Bi = \dfrac{hL}{k_{solid}} = \dfrac{\text{Conductive resistance in solid}}{\text{Convective resistance in the thermal boundary layer}}$

Bond number, $Bo = \dfrac{(\rho_l - \rho_v)gL^2}{\sigma} = \dfrac{\text{Body force}}{\text{Surface tension}}$

Capillary number, $Ca = \dfrac{We}{Re} = \dfrac{U\mu}{\sigma} = \dfrac{\text{Weber number}}{\text{Reynolds number}}$

Eckert number, $Ec = \dfrac{U^2}{c_p \Delta T} = \dfrac{\text{Kinetic energy in the flow}}{\text{Enthalpy difference in the boundary layer}}$

Fourier number, $Fo = \dfrac{\alpha \tau}{L^2} = \dfrac{\text{Rate of heat conduction}}{\text{Rate of thermal energy stored}}$

Galilei number, $Ga = \dfrac{gL^3}{\nu^2} = \dfrac{\text{Gravity force}}{\text{Viscous force}}$

Jakob number, $Ja = \dfrac{c_p(T_w - T_{sat})}{h_{fg}} = \dfrac{\text{Sensible heat}}{\text{Latent heat of this change}}$

Knudsen number, $Kn = \dfrac{\lambda}{L} = \dfrac{\text{Mean free path length}}{\text{Characteristic macroscopic length}}$

Mach number, $Ma = \dfrac{U}{a} = \dfrac{\text{Gas flow velocity}}{\text{Velocity of sound}}$

Marangoni number, $Mr = \dfrac{d\sigma}{dT} \cdot \dfrac{L\Delta T}{\mu\alpha} = \dfrac{\text{Thermal surface tension force}}{\text{Viscous force}}$

Nusselt number, $Nu = \dfrac{hL}{k_{fluid}} = \dfrac{\text{Convective heat transfer across the boundary layer}}{\text{Conductive heat transfer in the fluid}}$

Prandtl number, $Pr = \dfrac{\nu}{\alpha} = \dfrac{\text{Momentum diffusivity}}{\text{Thermal diffusivity}}$

Reynolds number, $Re = \dfrac{\rho UL}{\mu} = \dfrac{\text{Inertial force}}{\text{Viscous force}}$

Stanton number, $St = \dfrac{Nu}{RePr} = \dfrac{h}{\rho c_p U}$

Weber number, $We = \dfrac{\rho U^2 L}{\sigma} = \dfrac{\text{Inertial force}}{\text{Surface tension}}$

INDEX

Printed in the United States
By Bookmasters